WITHDRAWN

Principles of
Hydrocarbon Reservoir Simulation

Principles of Hydrocarbon Reservoir Simulation

G. W. Thomas

*Vice President, Research and Development
Scientific Software Corporation*

International Human Resources Development Corporation

Boston

Copyright © 1982 by International Human Resources Development Corporation. All rights reserved. No part of this book may be used or reproduced in any manner whatsoever without written permission of the publisher except in the case of brief quotations embodied in critical articles and reviews. For information address: IHRDC, Publishers, 137 Newbury Street, Boston, MA 02116.

ISBN: 0-934634-11-4

Library of Congress Catalog Card Number: 81-82011

Printed in the United States of America

Til
E.M.S.
og
S. T-S.
.
.
.

mitt
livs
lys

Preface

Who before dying demands not rebirth.

e.e. cummings

The material presented here constitutes an introduction to the basic principles involved in constructing a mathematical model to simulate hydrocarbon reservoir behavior. The point of view is that of a developer rather than a user. On the other hand, it is my firm belief that wise and efficient use of reservoir simulators requires a fundamental understanding of the physical and mathematical principles employed in their construction, and the shortcomings that devolve from the various approximations used. The book is primarily directed toward applied scientists, practicing engineers, and students of petroleum reservoir engineering. However, the basic ideas given here are also of use to hydrologists interested in simulating ground water flow.

Inasmuch as reservoir simulation technology is a rapidly expanding field, I make no claim to completeness. What is found here are the essential considerations assuming the ultimate goal is a three-phase multidimensional, black-oil simulator. These considerations should make the reader conversant with the major concerns of simulation technology, and equip him with the essential tools to understand much of the current literature on the topic. To this end, attention is devoted in the early chapters to the basic foundations of the subject from a descriptive, mathematical, and physical point of view.

Chapter 1 presents the essence of reservoir simulation without getting too technical. Chapter 2 is for the person who needs to brush-up on some basic mathematical concepts used throughout the text. Chapter 3 serves the purpose of review for the practicing reservoir engineer, and as a source of some reservoir engineering principles for those who come from outside the petroleum field. Chapter 4 sets forth, in a simple, consistent way, the mass conservation equations one confronts in treating flow in porous media. For the numerical analyst, chapter 5 can be glossed over. The material in this chapter, however, is essential for one who has no background in finite difference methods since subsequent chapters expand on ideas presented here. Chapter 6, while devoted to single-phase flow, contains fundamental material that is equally important in multiphase flow situations. Chapter 7 provides a taste of the current state of technology with respect to simulation of three-phase black-oil systems while chapter 8 treats the concepts of pseudo functions, handling of wells, gas percolation procedures, and variable bubble-point problems. A number of theorems and mathematical concepts important to developments throughout the book are included in an appendix for completeness.

To serve as a text, exercises are provided at the end of each chapter. Additional exercises will come to the mind of the instructor who reads with pencil in hand. It is strongly recommended that students be encouraged to develop computer programs to help them assimilate the theory and techniques presented here. The development of a 2-D or 3-D single-phase flow simulator is a reasonable goal toward the end of a semester's course. This provides an excellent vehicle on which to try the various direct and iterative solution procedures discussed here. Should the instruction continue beyond one semester, a rudimentary two-phase model can be beneficial in crystallizing the principles involved in simulating multiphase flow. Such a development can be executed as a group or class effort.

I stake no claims to originality, except for the errors that escaped undetected, and the style of presentation. With regard to the latter, I use the words "simulator" and "model" interchangeably with eyes closed to the distinction in meaning between the two. This sacrifice of purity is to avoid the tedium of repetition.

I owe much to the publications of others—particularly, D. W. Peaceman and K. H. Coats, whose developments extensively flavor parts of the book. There are other colleagues and associates in technical circles to whom I'm indebted. Among these, D. H. Thurnau, who willingly read the manuscript and granted me the benefits of his wise counsel. In no less measure, have I profited from association with students in the United States and abroad when acting in a former capacity as a university professor, or in my current one as a consultant.

This book had its birth in Norway in 1976, where it was first published as a compendium of lecture notes at the behest of students there. When it was no longer available, Michael Hays of International Human Resources Development Corporation encouraged its resurrection in revised form. Its current life is the result of his midwifery and the labors of the editorial staff of IHRDC. E. A. Breitenbach, Chief Executive Officer of Scientific Software Corporation, graciously gave his blessings to this undertaking. To these I express genuine appreciation.

May, 1981
Englewood, Colorado
G. W. Thomas

Table of Contents

Preface   vii

1. Overview of Reservoir Simulation   1
  1.1 Introduction
  1.2 Historical Perspective
  1.3 The Material Balance Equation
  1.4 Numerical Reservoir Simulators
  1.5 Types of Reservoir Simulators
  1.6 Use of a Reservoir Simulator
  1.7 Data Preparation
  1.8 History Matching and Performance Prediction
  1.9 References

2. Elementary Mathematical Concepts   11
  2.1 Introduction
  2.2 Elementary Vector Analysis
      2.2.1 Vector Gradient
      2.2.2 Vector Algebra
      2.2.3 Divergence
      2.2.4 Gauss' Theorem and the Continuity Equation
  2.3 Matrix Methods
      2.3.1 Matrices
      2.3.2 Types of Matrices
      2.3.3 Matrix Operations
      2.3.4 Determinants
      2.3.5 Matrix Inverse and Simultaneous Equations
      2.3.6 Matrix Eigenvalue Problem

- 2.4 Solution of Simultaneous Linear Algebraic Equations
  - 2.4.1 Gaussian Elimination
  - 2.4.2 Iterative Methods
- 2.5 Linear Algebra
  - 2.5.1 Definitions
- 2.6 Exercises
- 2.7 References

3. Properties of Reservoir Rocks and Fluids   33
- 3.1 Introduction
- 3.2 Basic Rock and Rock-Fluid Properties
  - 3.2.1 Permeability, Porosity, Saturation, and Compressibility
  - 3.2.2 Wettability
  - 3.2.3 Capillary Pressure
  - 3.2.4 Relative Permeability
- 3.3 Reservoir Fluid Properties
  - 3.3.1 Characteristics of Black-Oil Reservoir Fluids
  - 3.3.2 Some Characteristics of Compositional Fluids
- 3.4 Exercises
- 3.5 References

4. Reservoir Flow Equations   43
- 4.1 Introduction
- 4.2 Flow Potential
- 4.3 Darcy's Law
- 4.4 Equations of State
- 4.5 Recapitulation
- 4.6 Single-Phase Incompressible Flow
- 4.7 Single-Phase Compressible Flow
  - 4.7.1 Ideal Gas Flow
  - 4.7.2 Real Gas Flow
- 4.8 Multiphase Flow—The Generalized Flow Equation
- 4.9 Black-Oil Simulator
- 4.10 Exercises
- 4.11 References

5. Finite Difference Approximations   61
- 5.1 Introduction
- 5.2 Finite Differences
- 5.3 Application to Single-Phase Flow
  - 5.3.1 Explicit Method of Solution
  - 5.3.2 Crank-Nicolson Method
  - 5.3.3 Thomas' Algorithm
- 5.4 Stability Analysis
  - 5.4.1 Stability Analysis of the Crank-Nicolson Method
- 5.5 Truncation Error
- 5.6 Other Considerations
- 5.7 Exercises
- 5.8 References

6. Single-Phase Multidimensional Flow   79
- 6.1 Introduction
- 6.2 Grid Systems and Boundary Conditions

Table of Contents

    6.3 Slightly Compressible Flow
    6.4 Alternating Direction Implicit Procedure
    6.5 Iterative Alternating Direction Implicit Method
    6.6 Stability and Accuracy of ADI Methods
    6.7 Flow Problems in 3-D
    6.8 Band Matrix Problems
          6.8.1 Determination of Matrix Structures
    6.9 Ordering Schemes and Sparse Matrix Methods
    6.10 Strongly Implicit Procedure
    6.11 Point Iterative Methods
    6.12 Block Successive Overrelaxation
    6.13 Exercises
    6.14 References

7. Multiphase Flow   121
    7.1 Introduction
    7.2 Two-Phase Flow
          7.2.1 Simultaneous Solution Method
          7.2.2 Leap-Frog Technique
          7.2.3 IMPES Formulation
    7.3 Three-Phase Flow
          7.3.1 IMPES Formulation
          7.3.2 Stability of an IMPES Formulation
          7.3.3 Fully Implicit Formulation
          7.3.4 The Adaptive Implicit Method
    7.4 Exercises
    7.5 References

8. Special Concepts   153
    8.1 Introduction
    8.2 Treatment of Wells
          8.2.1 Production Rate Specified
          8.2.2 Specified Bottomhole Pressure
          8.2.3 Injection Wells
    8.3 Pseudo Functions
          8.3.1 Pseudo Functions Based on the Vertical Equilibrium Concept
          8.3.2 Gravity-Capillary Vertical Equilibrium
          8.3.3 Gravity-Segregated Vertical Equilibrium
          8.3.4 Validation of Vertical Equilibrium
          8.3.5 Dynamic Pseudo Functions
          8.3.6 Well Pseudo Functions
    8.4 Gas Percolation
    8.5 Variable Bubble-Point Problems
    8.6 Exercises
    8.7 References

Nomenclature   179

Appendix   185
A.1 Big "O" Notation
A.2 Functions of Class C^n
A.3 First Mean Value Theorem for Integrals

A.4 Leibniz' Rule for Integrals
A.5 Classification of Partial Differential Equations
    A.5.1 Linearity
    A.5.2 Canonical Forms
    A.5.3 Solution of Partial Differential Equations
A.6 Matrix Methods
    A.6.1 Rank of a Matrix
    A.6.2 Linear Independence of Eigenvectors
    A.6.3 Polynomials of a Matrix
    A.6.4 Gerschgorin's Theorem
    A.6.5 Brauer's Theorem
    A.6.6 Permutation Matrices
    A.6.7 Reducible Matrices
    A.6.8 Graph of a Matrix
    A.6.9 Irreducible Diagonal Dominance
    A.6.10 LU Decomposition
A.7 References

Index   201

*Of making many books there is no end
And much study is a weariness of the flesh.*

                                  Ecclesiastes

1
Overview of Reservoir Simulation

For out of olde feldes, as men seith,
Cometh al this newe corne fro yeere to yeere;
And out of olde bokes, in good feith
Cometh al this newe science that men lere.

Chaucer

1.1 Introduction

To give some perspective to our study and emphasize its importance, we briefly trace here the historical evolution within the petroleum industry that has led to current reservoir simulation technology. At the same time, we use this occasion to precisely define what we mean by a "model," especially a mathematical model. Finally, we discuss what is involved in constructing a reservoir simulator (without getting mathematical), types of reservoir simulators, how they are used, and problems involved in their use and misuse.

1.2 Historical Perspective

Models of one kind or another have been employed throughout the history of mankind. For the most part, they are used to obtain a better understanding of the environment and to predict the behavior of physical phenomena under the constraints of nature's laws. There has been an increasing dependence on models concurrent with the growth of reservoir engineering technology. This dependence is unique and borders on total commitment inasmuch as the environment we treat, the petroleum reservoir, is inaccessible in the large.

For our purposes, we consider a model as an entity permitting the study of phenomena, under appropriate test conditions, that are likely to occur in practice. In this context, a model can be a physical device wherein one attempts to reproduce in microcosm the desired phenomena. On the other hand, it can take the form of physical concepts, expressed mathematically, from which one derives his conclusions by appropriate mathematical techniques. We refer to this as a "mathematical model."

Both kinds of models have played an important role in the petroleum industry. For example, the laws governing fluid flow in porous media were discovered and delineated employing physical models. Darcy's Law, the concepts of relative permeability, capillary pressure, density, and viscosity correlations, and so on, all have their origins in experiments with physical models. Needless to say, they have been and are indispensable to the practice of reservoir engineering. Nevertheless, they have their limitations. These largely reside in the impracticality of rigorously modeling large scale systems such as an entire petroleum reservoir. Such an undertaking would require surmounting formidable problems at prohibitive costs, and even then, the desired goal of obtaining a generalized physical reservoir model may not be attained. What we are saying is that physical models are most useful in studying phenomena on a small scale, and indeed are essential to determine the physical concepts controlling these phenomena. When it comes to modeling global systems, such as petroleum reservoirs, we must appeal to a different approach, usually the mathematical one. The desire to adequately treat an entire reservoir with some degree of accuracy has given birth to the technology known as reservoir simulation. This is not to say that reservoir simulation techniques are limited to global situations. They are also used in studying local phenomena around wellbores, and have proven superior, in this regard, to physical models.

Possibly only the name "reservoir simulation" is new since the concepts involved have long been employed by reservoir engineers. Indeed, mathematical models, albeit simple ones, were devised in the days when reservoir engineering was still in its infancy. The most familiar of these is the material balance equation which we discuss in the next section. This is a mathematical model or reservoir simulator in every sense. It is based on a fundamental physical concept, namely, a conservation principle. This principle when expressed mathematically under the constraints of arbitrary assumptions constitutes the model. It is worthwhile to note that modern reservoir models are based on the same principles. They differ insofar as attempts have been made to lift the restricting assumptions inherent in the material balance equation and more closely approximate actual reservoir conditions.

1.3 The Material Balance Equation

It is not our purpose here to derive the material balance equation. Rather we seek to discuss the premise upon which it is based and illuminate its utility and limitations. This provides us with a convenient entrée for treating more advanced mathematical models.

The material balance equation was first introduced by Schilthuis[1] in 1936. He proposed treating a reservoir as a homogeneous tank having uniform rock and fluid properties throughout. Consequently, it is sometimes referred to as "the tank model." The tank is assumed sealed on all sides, i.e., it is a closed system with no flow across the boundaries. For this system, we invoke the conservation concept that the volume of fluids entering the tank, less the volume leaving, equals the net change in volume. We schematically depict this in Fig. 1.1. Since the tank is sealed, it is tacitly assumed the fluids are entering or leaving through wells (sources or sinks).

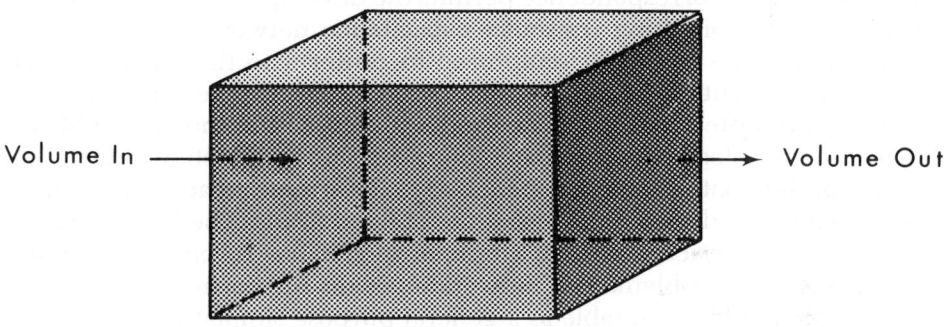

Fig. 1.1 Tank Model Concept.

The equation that evolves from this treatment, and its various modifications, has been an important tool to the reservoir engineer. It has made possible estimations of oil-in-place, gas-in-place, and the amount of water influx.[2] Furthermore, it has provided means to predict oil production under various driving mechanisms such as solution gas drive, gas cap drive, etc.[3] Another nice feature is that it yields a rather simple formula which an engineer can employ without resorting to a computer. Nevertheless, it has its drawbacks. These come into sharp relief when we compare the model to an actual reservoir.

First, the material balance equation (MBE) does not reflect the spatial variation of the rock and fluid parameters. Reservoirs actually are heterogeneous and anisotropic. Permeability, for example, changes from point to point (heterogeneity) and even at a given point, may take on different values depending upon the direction (anisotropy). Values of porosity and the phase behavior of fluids, their densities, etc., also can vary appreciably throughout the reservoir. Another major deficiency of the MBE is that the actual geometrical configuration of the reservoir is not considered. This can have important ramifications where fluid flow processes are strongly affected by the reservoir geometry. For example, rapid segregation of fluids in high relief reservoirs cannot be adequately modeled with the MBE. Furthermore, no provision is made to reflect the existence of wells and their locations within the system except to say that somehow fluids enter and fluids leave. Similarly, the dynamic effects of fluid movement within the reservoir are neglected, and it is not possible to ascertain the spatial distribution of water, oil and gas with time. Thus, as one produces a reservoir, we cannot predict the shifting of gas/oil or water/oil contacts. Nevertheless, we desire the capability to do this for obvious economic reasons.

The deficiencies of the MBE were recognized early and a number of means were taken to overcome them. Some of these exploit physical models more fully. One approach employs the similarity between flow of electricity and fluid flow. This correspondence permits the development of an electrical analog of a reservoir using resistance-capacitance networks.[4] Another approach uses a large container sculptured to conform to the boundaries of a reservoir. The container is filled with an aqueous electrolyte and individual wells are represented by copper electrodes placed within the system.[5] Current fed to the electrodes simulates injection and production rates. The objective of the "potentiometric model" is to determine steady-state potential distributions and reconstruct locations of flood fronts. The biggest weakness in these approaches is that a unique model has to be custom built for each reservoir problem. The network analyzer furthermore is cumbersome and not readily adaptable as a general purpose simulator. The potentiometric model is also restricted to fluid flow regimes having unit mobility ratios and cannot reflect reservoir heterogeneities and anisotropies. Since the advent of numerical methods these approaches are now rarely, if ever, used.

1.4 Numerical Reservoir Simulators

Before we discuss the principles involved in the construction of a numerical reservoir simulator, we first elucidate what we mean by a "numerical solution." Suppose we can express a particular physical process by a mathematical equation or collection of mathematical equations. We assume the process takes place in some region having finite dimensions. The equation(s) describing this process will invariably refer to a characteristic, arbitrary point within the region. By performing a pertinent mathematical process on the equation(s), we can determine the interaction of this point with all other points in the system at all times and thereby predict the behavior of the physical process under study. We refer to this procedure as solving the equation(s). If the mathematical equation(s) are not too complex, and we are sufficiently clever, our solution process may yield a formula which we can subsequently manipulate to compute the things we desire. Such a solution is called an *analytic solution*. If we're lucky, it can be achieved with nothing more than pure thought, some mathematical knowledge, pencil and paper. On the other hand, if our equation(s) are too complex to be solved analytically, then we must settle for something less than a formula to represent the solution. What we do in such cases, is replace the original equation(s) with a set of simpler ones that are within our capabilities to solve, and are related to the originals in some sense. However, rather than get a formula, we arrive at solutions to the simpler equations in the form of tables of numerical values each of which refers to discrete points in space and time within the region. We call this a *numerical solution*. As such, it represents an approximation to the original equation(s) we wanted to solve.

In our attempts to construct a mathematical model that overcomes the deficiencies of the MBE, we invariably arrive at equations that fall within the second category cited above; namely, those requiring numerical solutions, hence the name "numerical reservoir simulators." The volume of work required to achieve numerical answers to even the simplest reservoir problems is astronomical. Consequently, we must rely upon high speed digital computers to accomplish this task. Indeed, with the advent of high speed computing equipment, it is now possible to employ generalized reservoir simulators to study the behavior of any reservoir under a wide range of operating conditions.

To see what is involved in the construction of such a model, let us suppose we shrink the tank of the MBE down to a small element and consider it as one of many within the boundaries of the reservoir, each of which is contiguous with the others surrounding it. If we look at our reservoir in plan view, it would appear as in Fig. 1.2. Obviously, the smaller we shrink the elements or blocks, the more accurately we define the reservoir geometry. Now suppose we permit flow across the faces of the blocks interior to the reservoir, but over each of these blocks we now invoke the same conservation notion employed by Schilthuis,[1] however, in slightly different form, i.e., we state it as:

Rate of Fluid In − Rate of Fluid Out = Net Change in Fluid Rate

The collection of such material balance equations over each block constitutes the mathematical model.

The rate form has the nicety that we can employ Darcy's Law and introduce the dynamic effects of fluid movement. Furthermore, by segment-

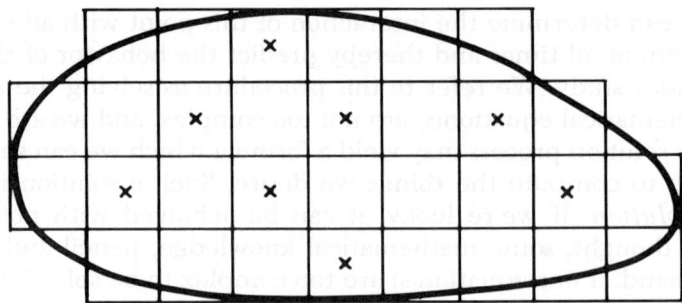

Fig. 1.2 Gridded Reservoir.

ing the reservoir into a collection of small blocks, we can assign unique values of the rock properties to each, and thereby approximate reservoir heterogeneities and anisotropies. Spatial variation of fluid properties can also be assigned blockwise or zonewise throughout the system. To reflect the existence of wells, we merely add appropriate source (for injection) or sink (for production) terms to the conservation equation for a given block in which the well occurs. Those blocks indicated by "x" in Fig. 1.2 are well blocks and the material balance expressions for these would be so modified. Because we permit flow across interior block boundaries, we can also track fluid front movements, monitor changes in gas/oil and water/oil contacts, and determine dynamic changes in pressure and saturation distributions. In brief, this approach essentially removes all the shortcomings inherent in the MBE.

Because properties do not vary from point to point within the simple tank model, it is sometimes referred to as the *zero-dimensional model*. In our more realistic approach, we can take two or more such tanks butted end-to-end, assign unique rock and/or fluid properties to each, and permit flow from one block to the other through the adjoining faces. This constitutes a one-dimensional (1-D) model. Similarly, this procedure can be extended to reservoirs where the property variations and flow are permitted over two and three dimensions giving rise to 2-D and 3-D simulators.

1.5 Types of Reservoir Simulators

Reservoir simulators can be classified according to the type of reservoir they are intended to simulate or on the basis of a particular reservoir process.[6] Simulators based on the former classification fall generally into three groups: gas reservoir simulators, black-oil reservoir simulators, and compositional reservoir simulators. Gas reservoir simulators may be either single-phase or two-phase models depending on whether or not mobile water is present. A black-oil model, on the other hand, is capable of simulating those systems where gas, oil, and water are present in all proportions. Usually phase transfers are accounted for between the gas and oil phases, i.e., they include the effects of gas going in and out of solution in the oil. Condensate and volatile oil reservoirs usually require a special purpose simulator that accounts for the compositional behavior between the individual hydrocarbon components in the gas and liquid phases. This type of model is among the most complex since it focuses on individual components rather than phases. Its operation is also more costly than a black-oil simulator.

Particular reservoir processes and phenomena such as wellbore coning, thermal recovery processes, chemical flooding, and miscible displacement categorize other types of reservoir models. In coning models, single-well behavior is examined in detail with the purpose of determining the best completion intervals and production rates necessary to minimize gas or water coning into a well.[7] Thermal recovery processes, including steam stimulation, steam displacement and underground combustion, have given rise to very sophisticated reservoir models that attempt to account for all the pertinent physical phenomena involved.[8-10]

Chemical flood models are characterized by additional conservation equations for various species, e.g., polymers, surfactants, etc.[11] In addition, they must have some representation of adsorption, and include the effects of permeability reduction in the aqueous phase after contact with the polymer. Most miscible displacement models today rely upon modifications of models used for immiscible displacement processes.[12] For the most part, these are capable of simulating the essential features of miscible displacement; however, they are not capable of simulating the fine structure of unstable miscible flow.[13]

1.6 Use of a Reservoir Simulator

In conducting a reservoir simulation study, a number of decisions have to be made with regard to model selection. Significant parameters that enter into this decision are the reservoir type, reservoir geometry and dimensionality, the availability of data, type of reservoir process being simulated, computer availability and manpower requirements. The reservoir type of process will, of course, dictate whether to choose a gas model, a black-oil model, or a special purpose model, etc. In many applications, 1-D models are adequate. For example, the high pinnacle reef structures of Alberta where the dominant flow direction is vertical are candidates for 1-D simulation. When the effects of areal changes are important, then a 2-D model is best suited to the purpose. Two-dimensional models are also useful in studying reservoir behavior in cross section. For example, if one wishes to examine the effects of gravity segregation and crossflow, then a 2-D cross-sectional model would be a proper choice. Simulation of reservoir phenomena in 3-D is usually a costly proposition. If the reservoir thickness is large in relation to its areal extent or if there is pronounced heterogeneity in the vertical direction, then usually 3-D simulation is necessary.

The data requirements increase with the complexity of the model. Consequently, the model choice should be consistent with the minimum data required to adequately define the reservoir. If adequate field data exist, reasonably accurate performance predictions can be made. If data are incomplete or suspect, simulators may be used to compare qualitatively the results of different ways of operating the reservoir. In either case, the accuracy of the simulator can be improved by adjusting data using a technique known as history matching.

1.7 Data Preparation

After the type of model has been selected, the reservoir is divided into a number of cells or blocks. Each cell is identified by its x, y or z-coordinates or most often, by its i,j,k-indices, where, for uniform grids, $x = i\Delta x$, $y = j\Delta y$, and $z = k\Delta z$. Normally, the reservoir is considered sealed on its exterior boundaries although efflux or influx at an assigned pressure or rate can be specified.

To each block or cell, we assign the rock properties, geometrical data and well data (if the cell is a well block). The rock data are specific permeability and porosity. The geometrical data consist of the cell dimensions, i.e., Δx, Δy, and thickness Δz (or h), and the cell elevation relative to some datum. If a well falls in a block then pertinent well data must be included such as well type, rates, completion intervals, etc. In some instances, though not always, one may also specify pressures and phase saturations initially existing in each block. Relative permeability, capillary pressure and PVT data must also be supplied. Frequently, different sets of these data are specified for different regions of the reservoir as required for a specific problem.

1.8 History Matching and Performance Prediction

The first step in a history match is to calculate reservoir performance using the best data available. The results are compared with the field recorded histories of the wells. If the agreement is not satisfactory, such data as permeability, relative permeability, and porosity are varied from one computer run to another until a match is achieved. The simulator is then used to predict performance for alternative plans of operating the reservoir.

The behavior of the reservoir is influenced by many factors—permeability, porosity, thickness, saturation distributions, relative permeability, etc., that are never known precisely in the system. What the engineer determines is a combination of these variables, which results in a match. This combination is not unique, so it may not precisely represent reservoir conditions. When the simulator, after a match, is used to predict, it is not certain that the physical picture of the reservoir described in the simulator will give predictions sufficiently close to actual reservoir performance. In general, the longer the matched history period, the more reliable the predicted performance will be. It behooves the engineer to monitor periodically the predicted versus the actual performance and update his physical picture of the reservoir.

We emphasize that the central purpose of reservoir simulation is to evaluate field performance under a variety of operating policies. The idea is to determine the set of producing conditions that will maximize recoveries at the least cost. To do this with a simulator requires using the best data possible and exercising good engineering judgment.

1.9 References

1. Schilthuis, R.J.: "Active Oil and Reservoir Energy," *Trans.*, AIME (1936) **118**, 33.
2. Craft, B.C. and Hawkins, M.F.: *Applied Petroleum Reservoir Engineering*, Prentice-Hall, Inc., Englewood Cliffs (1959).
3. Pirson, S.J.: *Elements of Oil Reservoir Engineering*, second edition, McGraw-Hill Book Co. Inc., New York City (1958) 2.
4. Bruce, W.A.: "An Electrical Device for Analyzing Oil Reservoir Behavior," *Trans.*, AIME (1943) **146**, 112.
5. Lee, B.D.: "Potentiometric Model Studies of Fluid Flow in Petroleum Reservoirs," *Trans.*, AIME (1948) **174**, 41.
6. Odeh, A.S.: "Reservoir Simulation—What Is It?," *J. Pet. Tech.* (Nov. 1969) 1383.
7. MacDonald, R.C. and Coats, K.H.: "Methods for Numerical Simulation of Water and Gas Coning," *Trans.*, AIME (1970) **249**, 423.
8. Crookston, R.B., Culham, W.E., and Chen, W.H.: "Numerical Simulation Model for Thermal Recovery Processes," *Soc. Pet. Eng. J.* (Feb. 1979) 37.
9. Grabowski, J.W., Vinsome, R.K., Lin, R.C., Behie, A., and Rubin, B.: "A Fully Implicit General Purpose Finite-Difference Thermal Model for In Situ Combustion and Steam," paper SPE 8396 presented at the 54th Annual Technical Conference and Exhibition, Las Vegas, Sept. 23–26, 1979.
10. Coats, K.H.: "A Fully Implicit Steamflood Model," *Soc. Pet. Eng. J.* (Oct. 1978) 369.

11. Limon, J.T., Thomas, G.W., and Zetik, D.F.: "A Numerical Investigation of Micellar Flooding," *J. Can. Pet. Tech.* (July–Sept. 1980) 111.
12. Lantz, R.B.: "Rigorous Calculation of Miscible Displacement Using Immiscible Reservoir Simulators," *Soc. Pet. Eng. J.* (June 1970) 192.
13. Todd, M.R. and Longstaff, W.J.: "The Development, Testing, and Application of a Numerical Simulator for Predicting Miscible Flood Performance," *J. Pet. Tech.* (July 1972) 894.

2
Elementary Mathematical Concepts

*Suppose you've no direction in you,
I don't see but you must continue
To use the gift you do possess,
And sway with reason more or less.*

Robert Frost

*Mazes intricate
Eccentric, interwov'd, yet regular
Then most, when most irregular they seem.*

John Milton

2.1 Introduction

To adequately grasp the underlying concepts involved in the structure, development and use of a reservoir simulator, it is necessary that we have an understanding of some basic mathematical tools. In particular, we employ some elementary ideas derived from vector analysis and matrix theory. In what follows, we review only those basic concepts that are essential to our purpose. Furthermore, the material presented here is restricted to the premise that solutions to reservoir engineering problems are to be achieved by numerical means.

2.2 Elementary Vector Analysis[1]

We recall that a vector is usually defined as a directed line segment. It is a quantity having both magnitude and direction. If we denote a vector by v, then its magnitude is denoted by |v|, also called the *modulus* or *norm* of v. Thus, the velocity of a particle of fluid at a point P in a reservoir R is a vector. It is differentiated from the scalar quantities, temperature and density at that point, in that the latter are not characterized by a directional property.

Consider a cartesian coordinate system consisting of the x, y and z

axes. If a vector resides in this 3-dimensional space, then we can construct its vector components by simply making projections on each of the coordinate axes. Vector **v** is then the resultant of its vector components, i.e.,

$$\mathbf{v} = \mathbf{v}_1 + \mathbf{v}_2 + \mathbf{v}_3. \tag{2.1}$$

Eq. 2.1 can be further expressed in terms of "unit vectors" **i**, **j**, **k** each with moduli unity having directions that parallel the x, y and z coordinate

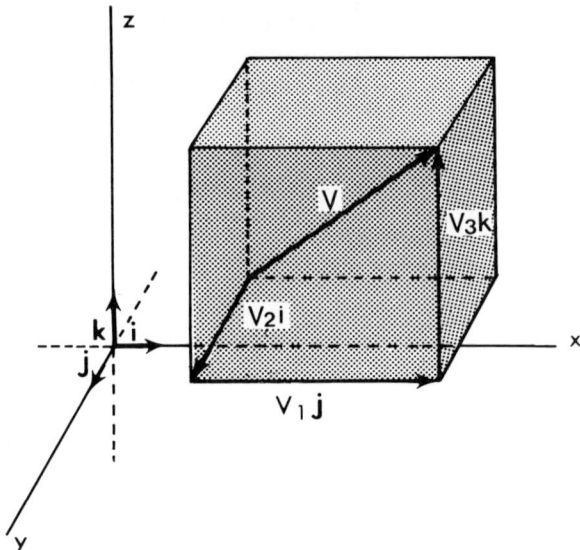

Fig. 2.1 Cartesian Coordinates.

axes, respectively (see Fig. 2.1). Consequently, if $|\mathbf{v}_1| = v_1$, $|\mathbf{v}_2| = v_2$, $|\mathbf{v}_3| = v_3$ then

$$\mathbf{v} = v_1 \mathbf{i} + v_2 \mathbf{j} + v_3 \mathbf{k} \tag{2.2}$$

where, v_1, v_2, v_3 are the "scalar components" of **v**. Now suppose we shift the position of **v** along the ray extending from the origin in the direction of **v** ; i.e., we move **v** either closer to or farther from the origin without changing its direction. Regardless of the position **v** occupies on this ray, its representation is still given by Eq. 2.2. Stated another way, **v** is uniquely determined by its scalar components. This observation makes possible an alternative definition of a vector in 3-space; namely, a vector is simply an ordered triple of numbers, i.e.,

$$\mathbf{v} = (v_1, v_2, v_3). \tag{2.3}$$

We refer to the collection of all such vectors as a 3-dimensional Euclidean space, E_3, if Eqs. 2.6–2.9 (See section 2.2.2) are satisfied.

From the mathematical point of view, there is no reason to confine ourselves to 3-dimensional Euclidean spaces. We can generalize the concept illustrated above and simply say that a vector is an ordered n-tuple of numbers $n \geq 1$, and the collection of all such vectors constitutes an n-dimensional Euclidean space, E_n†. In adopting this definition, we recognize that the concept of an n-dimensional vector (where $n > 3$) is an abstract one. It is not possible to display it pictorially. Furthermore, the dimensions of a Euclidean space should not be confused with the dimensionality of a reservoir which is modeled at most in 3-dimensions. Stated another way, a Euclidean space is not the spatial configuration we assign to a reservoir. Rather it is a mathematical entity that provides us a framework within which we can discuss the numerical solution of reservoir engineering problems. Thus we shall see, that the computation of pressure at n-ordered points in a reservoir can be thought of as finding the vector $(p_1, p_2, p_3, \ldots, p_n)$. In this case, we speak of such a set of points as the "solution vector."

2.2.1 Vector Gradient

Let Φ be a scalar function such that $\frac{\partial \Phi}{\partial x}, \frac{\partial \Phi}{\partial y}, \frac{\partial \Phi}{\partial z}$ are continuous at some point P in R. Physically these represent rates of change with respect to distance in each of the coordinate directions x, y and z. The gradient of Φ is given by

$$\text{grad } \Phi = \frac{\partial \Phi}{\partial x}\mathbf{i} + \frac{\partial \Phi}{\partial y}\mathbf{j} + \frac{\partial \Phi}{\partial z}\mathbf{k}. \tag{2.4}$$

The "del" or "nabla" operator is frequently used as a differential operator

$$\nabla \equiv \frac{\partial}{\partial x}\mathbf{i} + \frac{\partial}{\partial y}\mathbf{j} + \frac{\partial}{\partial z}\mathbf{k};$$

thus, grad $\Phi = \nabla \Phi$.

Suppose $\Phi(x, y, z) = c$ represents surfaces, S_c, in R for all values of the constant c. We refer to the scalar field Φ whose gradient is $\nabla \Phi$ as the *potential* of the vector field $\nabla \Phi$. The corresponding surfaces, S_c, are *equipotential surfaces*. One of the problems we are confronted with in simulating reservoir behavior is determining potential distributions, or the potential

---

† If v_i, $i = 1, 2, \ldots, n$ are real then E_n is a real Euclidean n-space. If they are complex, then E_n is a complex Euclidean n-space.

gradients $\nabla\Phi$, throughout the system. The potential gradient $\nabla\Phi$ has the following important properties:

(1) It is a vector function.
(2) Its direction is in the direction of maximum increase of Φ.
(3) It is always perpendicular to the equipotential surface, S_c, defined by $\Phi(x, y, z) = c$.
(4) It remains invariant under a coordinate transformation.

2.2.2 Vector Algebra

At this point it is desirable to consider algebraic manipulations of vectors. Consistent with our desire to only touch on those aspects needful to our ultimate purposes, we make no effort to completely cover vector algebra.

Vector addition and subtraction are defined on a component basis. If

$$\mathbf{a} = (a_1, a_2, \ldots, a_n) \text{ and } \mathbf{b} = (b_1, b_2, \ldots, b_n),$$

then we define

$$\mathbf{a} \pm \mathbf{b} = (a_1 \pm b_1, a_2 \pm b_2, \ldots, a_n \pm b_n) \tag{2.5}$$

where the only requirement we impose on **a** and **b** is that they both have the same number of components. In such a case, we say they are *conformable for addition*. The following is also true:

$$\mathbf{a} + \mathbf{b} = \mathbf{b} + \mathbf{a} \quad \text{(commutativity)} \tag{2.6}$$

$$\mathbf{a} + (\mathbf{b} + \mathbf{c}) = (\mathbf{a} + \mathbf{b}) + \mathbf{c} \quad \text{(associativity)} \tag{2.7}$$

$$\mathbf{a} + (-\mathbf{a}) = \mathbf{0} \tag{2.8}$$

$$\mathbf{a} + \mathbf{0} = \mathbf{a} \tag{2.9}$$

The vector, **0**, is the null vector consisting of zero elements.

We consider also scalar multiplication of a vector. If α is a scalar and $\mathbf{a} = (a_1, a_2, \ldots, a_n)$ then $\alpha\mathbf{a} = (\alpha a_1, \alpha a_2, \ldots, \alpha a_n)$. Furthermore, $\alpha\mathbf{a} = \mathbf{a}\alpha$, $\alpha(\mathbf{a} + \mathbf{b}) = \alpha\mathbf{a} + \alpha\mathbf{b}$ and $(\alpha + \beta)\mathbf{a} = \alpha\mathbf{a} + \beta\mathbf{a}$ for β another scalar. Moreover, $(\alpha\beta)\mathbf{a} = \alpha(\beta\mathbf{a})$ and $1\mathbf{a} = \mathbf{a}$. All vectors in a Euclidian space satisfying these properties of scalar multiplication constitute a *vector space*.

The *dot or inner product* of two vectors **a** and **b** is given by

$$\mathbf{a} \cdot \mathbf{b} = (a_1 b_1 + a_2 b_2 + \ldots + a_n b_n). \tag{2.10}$$

Elementary Mathematical Concepts

It can also be shown that

$$\mathbf{a} \cdot \mathbf{b} = |\mathbf{a}| |\mathbf{b}| \cos \theta \tag{2.11}$$

where θ is the angle between $\mathbf{a}$ and $\mathbf{b}$. If $\mathbf{a} \cdot \mathbf{b} = 0$ then we say $\mathbf{a}$ and $\mathbf{b}$ are orthogonal. Furthermore, the length or norm of a vector is given by

$$|\mathbf{a}| = (\mathbf{a} \cdot \mathbf{a})^{1/2}.$$

The dot product satisfies the following properties:

$$\mathbf{a} \cdot \mathbf{b} = \mathbf{b} \cdot \mathbf{a} \quad \text{(commutativity)} \tag{2.12}$$

$$\mathbf{a} \cdot (\mathbf{b} + \mathbf{c}) = \mathbf{a} \cdot \mathbf{b} + \mathbf{a} \cdot \mathbf{c} \quad \text{(distributive)} \tag{2.13}$$

Since the cartesian unit vectors $\mathbf{i}$, $\mathbf{j}$, $\mathbf{k}$ are at right angles to each other (i.e., they are mutually orthogonal) then it follows that $\mathbf{i} \cdot \mathbf{i} = \mathbf{j} \cdot \mathbf{j} = \mathbf{k} \cdot \mathbf{k} = 1$ and $\mathbf{i} \cdot \mathbf{j} = \mathbf{i} \cdot \mathbf{k} = \mathbf{j} \cdot \mathbf{k} = 0$.

In Eq. 2.11 we can consider the quantity $|\mathbf{a}| \cos \theta$ as the component of vector $\mathbf{a}$ in the direction of $\mathbf{b}$. This interpretation is useful in deriving the continuity equation using vector analysis principles as will be seen. It is important to notice that forming dot products of vectors results in a scalar quantity. Thus, it is sometimes called the *scalar product*.

Another kind of product, the *cross-product* or *vector product* can be defined such that a vector rather than a scalar quantity is obtained. It is particularly useful in depicting those processes characterized by rotational flow. Since such regimes are generally negligible in global reservoir problems, we make no attempt to treat it here. The interested reader is referred to any elementary text on vector analysis for further details.

2.2.3 Divergence

Let $\mathbf{v}(x, y, z)$ be a velocity vector at a point P in 3-space. The divergence of $\mathbf{v}$ is given by

$$\text{div } \mathbf{v} = \nabla \cdot \mathbf{v} = \left(\frac{\partial}{\partial x} \mathbf{i} + \frac{\partial}{\partial y} \mathbf{j} + \frac{\partial}{\partial z} \mathbf{k} \right) \cdot (v_1 \mathbf{i} + v_2 \mathbf{j} + v_3 \mathbf{k}), \tag{2.14}$$

$$\therefore \text{div } \mathbf{v} = \frac{\partial v_1}{\partial x} + \frac{\partial v_2}{\partial y} + \frac{\partial v_3}{\partial z}. \tag{2.15}$$

We observe the following facts about the divergence of a vector:

(1) It is a scalar quantity.
(2) It remains invariant under a coordinate transformation.

(3) If ρ is the density of a fluid at P having a velocity v, then $\rho v = q$ will be the *mass flux* at P. Physically, div q represents the rate of decrease of mass per unit volume in the neighborhood of the point P where div v is defined.

At this point, it is worthwhile to emphasize a common property of both the gradient of a scalar quantity and the divergence of a vector: both remain unchanged when the system of coordinates is altered. In other words, they are independent of the coordinate system we desire to employ. This means that if we can express the equations describing fluid flow in a reservoir exclusively in terms of these vector quantities, they will be universally applicable regardless of the coordinate system we impose on the system. The key to doing this is to employ Gauss' theorem which is stated without proof in the next section. Before treating that, it is of interest to consider the divergence of the gradient of Φ, i.e., div (grad Φ) = $\nabla \cdot \nabla \Phi$.

$$\nabla \cdot \nabla \Phi = \left(\frac{\partial}{\partial x} i + \frac{\partial}{\partial y} j + \frac{\partial}{\partial t} k \right) \cdot \left(\frac{\partial \Phi}{\partial x} i + \frac{\partial \Phi}{\partial y} j + \frac{\partial \Phi}{\partial t} k \right) \quad (2.16)$$

$$= \frac{\partial^2 \Phi}{\partial x^2} + \frac{\partial^2 \Phi}{\partial y^2} + \frac{\partial^2 \Phi}{\partial z^2} \equiv \nabla^2 \Phi \quad (2.17)$$

where

$$\nabla^2 = \frac{\partial^2}{\partial x^2} + \frac{\partial^2}{\partial y^2} + \frac{\partial^2}{\partial z^2}, \text{ the Laplacian operator.}$$

2.2.4 Gauss' Theorem and the Continuity Equation

Gauss' theorem (also called the Divergence Theorem) relates an integral over a volume, R, to an integral defined on its surface, S, namely,

$$\int_R \text{div } v \, dv = \int_S v \cdot d\sigma = \int_S v \cdot n \, ds \quad (2.18)$$

where v is a velocity vector in R, dv is a differential element of volume in R, $d\sigma$ is a directed element of surface $= n \, ds$, and n is an outward drawn unit vector normal to the scalar surface element, ds as depicted in Fig. 2.2. If we consider the fluid flux $q = \rho v$ at a point P then

$$\int_R \text{div } (\rho v) \, dv = \int_S \rho v \cdot n \, ds. \quad (2.19)$$

Elementary Mathematical Concepts

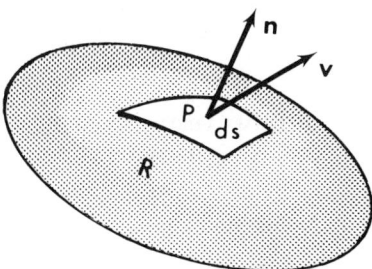

Fig. 2.2 Element of Surface in R.

Now since $\rho \mathbf{v} \cdot \mathbf{n}\, ds = |\rho \mathbf{v}|\, |\mathbf{n}\, ds| \cos \theta = \rho\, ds\, |\mathbf{v}| \cos \theta$ where θ is the angle between vectors $\mathbf{n}$ and $\mathbf{v}$, then $\rho \mathbf{v} \cdot \mathbf{n}\, ds$ physically represents the component of the fluid flux escaping from R through the element of surface ds in the direction of the outward drawn normal. Consequently, the integral of this quantity over the entire surface of R, i.e., the right-hand side of Eq. 2.19 represents the rate of decrease of mass from R. This can also be expressed as

$$\int_R -\frac{\partial}{\partial t}(\phi \rho)\, dv$$

where ϕ is the porosity. Therefore it follows that

$$\int_S \rho \mathbf{v} \cdot \mathbf{n}\, ds = -\int_R \frac{\partial}{\partial t}(\phi \rho)\, dv \qquad (2.20)$$

or combining Eq. 2.19 and Eq. 2.20

$$\int_R \operatorname{div}(\rho \mathbf{v})\, dv = -\int_R \frac{\partial}{\partial t}(\phi \rho)\, dv. \qquad (2.21)$$

Since R is an arbitrary volume, it follows that the arguments of the integrals in Eq. 2.21 are identical, i.e.,

$$\operatorname{div}(\rho \mathbf{v}) = -\frac{\partial}{\partial t}(\phi \rho). \qquad (2.22)$$

Eq. 2.22 is known as the *continuity equation*. It simply is an expression of the law of conservation of mass at a point P in R.

If a source or sink is present at the point P, then we add a mass rate term, $\tilde{g}$ say, to the continuity equation,

$$\nabla \cdot (\rho v) \pm \tilde{g} = -\frac{\partial}{\partial t}(\phi \rho) \cdot \qquad (2.23)$$

The choice of sign on the additive term is purely arbitrary. We adopt the convention that the minus sign represents a source and the plus sign a sink.

2.3 Matrix Methods[2]

2.3.1 Matrices

A matrix is simply a rectangular array of elements arranged in horizontal rows and vertical columns. Thus,

$$A = \begin{bmatrix} 1 & 2 & -3 \\ 4 & 0 & 5 \\ 0 & 1 & 4 \end{bmatrix}, \quad B = \begin{bmatrix} 1 & 2 \\ 0 & 1 \\ -2 & 4 \end{bmatrix}, \quad C = \begin{bmatrix} \alpha & \zeta & \kappa \\ \delta & \epsilon & \rho \end{bmatrix}$$

are examples of matrices. We say a matrix is of order $m \times n$ if it consists of m rows and n columns. If $m = n$, we call the matrix a square matrix of nth order. Consequently, A is a 3rd order square matrix, B is 3×2, and C is 2×3. In general, an $m \times n$ matrix will be denoted by

$$A = \begin{bmatrix} a_{11} & a_{12} & \cdots & a_{1n} \\ a_{21} & a_{22} & \cdots & a_{2n} \\ \vdots & \vdots & & \vdots \\ a_{m1} & a_{m2} & & a_{mn} \end{bmatrix}.$$

This can be conveniently abbreviated by $A = [a_{ij}]$ meaning A is a collection of elements with row index i and column index j.

The elements of a matrix need not be numbers. They can be functions, operators or even other matrices. We emphasize that a matrix is not a number nor can it be evaluated (like a determinant) to yield a number.

2.3.2 Types of Matrices

For our purposes, we will be working with square matrices almost exclusively, i.e., matrices of the form

$$A = \begin{bmatrix} a_{11} & a_{12} & \cdots & a_{1n} \\ a_{21} & a_{22} & \cdots & a_{2n} \\ \vdots & \vdots & & \vdots \\ a_{n1} & a_{n2} & & a_{nn} \end{bmatrix}.$$

The collection of elements a_{ii} is called the main diagonal of the matrix. If all the elements of **A** are zero except possibly those on the main diagonal then **A** is called a *diagonal matrix*. This is conveniently depicted by

$$\mathbf{A} = \begin{bmatrix} a_{11} & & 0 \\ & \ddots & \\ 0 & & a_{nn} \end{bmatrix}.$$

If $a_{ii} = \alpha$, a constant for all i then **A** is called a *scalar matrix*. An important scalar matrix, called the identity matrix, arises when $\alpha = 1$, and is symbolized by **I**.

A lower triangular matrix, **L**, is a square matrix where $a_{ij} = 0$ for $i < j$ while an upper triangular matrix, **U**, has elements $a_{ij} = 0$ for $i > j$. We will have occasion to consider the transpose of a matrix, i.e., if $\mathbf{A} = [a_{ij}]$ then the transpose of **A** denoted by $\mathbf{A}^T$ is given by $\mathbf{A}^T = [a_{ji}]$. Notice to form $\mathbf{A}^T$ we merely interchange rows and columns of **A**. Thus for

$$\mathbf{A} = \begin{bmatrix} a_{11} & a_{12} & \cdots & a_{1n} \\ a_{21} & a_{22} & \cdots & a_{2n} \\ \vdots & \vdots & & \vdots \\ a_{n1} & a_{n2} & & a_{nn} \end{bmatrix} \qquad \mathbf{A}^T = \begin{bmatrix} a_{11} & a_{21} & \cdots & a_{n1} \\ a_{12} & a_{22} & & a_{n2} \\ \vdots & \vdots & & \vdots \\ a_{1n} & a_{2n} & & a_{nn} \end{bmatrix}.$$

A square matrix **A** is said to be *symmetric* if $\mathbf{A} = \mathbf{A}^T$. If $\mathbf{A} = -\mathbf{A}^T$ then it is *skew symmetric*. It will be observed that the vector $\mathbf{v} = (v_1, v_2, \ldots, v_n)$ is a $1 \times n$ matrix, i.e., it is a *row matrix*. Similarly $\mathbf{v}^T$, its transpose,

$$\mathbf{v}^T = \begin{bmatrix} v_1 \\ v_2 \\ \vdots \\ v_n \end{bmatrix}$$

is an $n \times 1$ *column matrix*. Furthermore, we refer to the rows and columns of an $n \times n$ matrix **A** as the *row vectors* and *column vectors*, respectively since each is an ordered n-tuple of elements.

2.3.3 Matrix Operations

Below we outline the essential operations involving matrices. These involve equality, addition and subtraction, scalar multiplication and matrix multiplication.

Two matrices **A** and **B** are equal if and only if they are of the same order and $a_{ij} = b_{ij}$ for every i, j. If $\mathbf{A} = [a_{ij}]$ and $\mathbf{B} = [b_{ij}]$ and **A** and **B** are of the same order, then we can write $\mathbf{C} = \mathbf{A} \pm \mathbf{B}$ where $c_{ij} = a_{ij} \pm b_{ij}$ for every i, j. If α is a scalar and $\mathbf{A} = [a_{ij}]$ then we can say $\mathbf{B} = \alpha \mathbf{A}$ where $b_{ij} = \alpha a_{ij}$ for every i, j. The following properties hold for these operations:

$$\mathbf{A} + \mathbf{B} = \mathbf{B} + \mathbf{A}$$
$$\mathbf{A} + (\mathbf{B} + \mathbf{C}) = (\mathbf{A} + \mathbf{B}) + \mathbf{C}$$
$$\mathbf{A} + \mathbf{0} = \mathbf{A}$$
$$\alpha \mathbf{A} = \mathbf{A} \alpha$$
$$\alpha(\mathbf{A} + \mathbf{B}) = \alpha \mathbf{A} + \alpha \mathbf{B}$$
$$(\alpha_1 + \alpha_2)\mathbf{A} = \alpha_1 \mathbf{A} + \alpha_2 \mathbf{A}$$
$$(\alpha_1 \alpha_2)\mathbf{A} = \alpha_1(\alpha_2 \mathbf{A})$$

The matrix, **0**, is the zero or null matrix having all entries zero.

We would also like to have a definition of matrix multiplication; i.e., if $\mathbf{A} = [a_{ij}]$ and $\mathbf{B} = [b_{ij}]$ then we consider the product $\mathbf{AB}$. If $\mathbf{A}$ is $m \times n$ and $\mathbf{B}$ is $n \times p$ then for $\mathbf{C} = \mathbf{AB}$ we define

$$c_{ij} = \sum_{k=1}^{n} a_{ik} b_{kj}, \quad \begin{matrix} i = 1, 2, \ldots, m \\ j = 1, 2, \ldots, p. \end{matrix} \qquad (2.24)$$

The important requirement for matrix multiplication is that the number of columns of **A** must equal the number of rows of **B**. If this is satisfied we say **A** and **B** are *comformable for multiplication*. The resultant matrix **C** will be of order $m \times p$. Thus, if **A** is 4×3 and **B** is 3×5, **C** will be 4×5. On the other hand, if **A** is 6×4 and **B** is 3×6, **AB** is undefined, however, **BA** will be defined since $(3 \times \overset{*}{6})$ $(\overset{*}{6} \times 4)$ has adjacent numbers (indicated by asterisks) equal. In general, one may determine conformability for multiplication by considering the expression $(m \times \overset{*}{n})$ $(\overset{*}{n} \times p)$. If the starred quantities are equal, then multiplication is defined. We consider an example: Find **AB** and **BA** if

$$\mathbf{A} = \begin{bmatrix} 2 & 1 \\ -1 & 3 \end{bmatrix} \text{ and } \mathbf{B} = \begin{bmatrix} 4 & 0 \\ 1 & 2 \end{bmatrix}.$$

$$\mathbf{AB} = \begin{bmatrix} 2 & 1 \\ -1 & 3 \end{bmatrix} \begin{bmatrix} 4 & 0 \\ 1 & 2 \end{bmatrix} = \begin{bmatrix} 2(4) + 1(1) & 2(0) + 1(2) \\ -1(4) + 3(1) & -1(0) + 3(2) \end{bmatrix},$$

$$\therefore \mathbf{AB} = \begin{bmatrix} 9 & 2 \\ -1 & 6 \end{bmatrix}.$$

$$\mathbf{BA} = \begin{bmatrix} 4 & 0 \\ 1 & 2 \end{bmatrix} \begin{bmatrix} 2 & 1 \\ -1 & 3 \end{bmatrix} = \begin{bmatrix} 4(2) + 0(-1) & 4(1) + 0(3) \\ 1(2) + 2(-1) & 1(1) + 2(3) \end{bmatrix},$$

$$\therefore \mathbf{BA} = \begin{bmatrix} 8 & 4 \\ 0 & 7 \end{bmatrix}.$$

Notice both products **AB** and **BA** are defined, but $\mathbf{AB} \neq \mathbf{BA}$. In general, matrix multiplication is not commutative. However, matrix multiplication satisfies the following properties:

$$(AB)C = A(BC)$$

$$A(B + C) = AB + AC$$

$$(B + C)A = BA + CA$$

We know that if x and y are real numbers then $xy = 0$ implies that either $x = 0$ or $y = 0$ or both. Matrices, however, do not possess this property. Furthermore, if $AB = AC$ then this does not imply that $B = C$. Consequently, cancellation or division is not a valid operation in matrix algebra.

2.3.4 Determinants

A determinant is a single number that we associate with a square matrix **A**. It is denoted by

$$\det(\mathbf{A}) = \begin{vmatrix} a_{11} & a_{12} & \ldots & a_{1n} \\ a_{21} & a_{22} & & a_{2n} \\ \vdots & \vdots & & \vdots \\ a_{n1} & a_{n2} & & a_{nn} \end{vmatrix} = |\mathbf{A}|.$$

It should not be confused with the matrix itself. To compute $\det(\mathbf{A})$ we consider minors and cofactors. Given a square matrix **A**, a *minor* is the determinant of any square submatrix of **A** obtained by removal of an equal number of rows and columns. The *cofactor* of the element a_{ij} is a scalar obtained by multiplying together the term $(-1)^{i+j}$ and the minor M_{ij} obtained by removing the ith row and the jth column. To find $\det(\mathbf{A})$ we proceed as follows: (1) Select any row or column of **A** (say the kth row or column); (2) for each element in this row or column find the cofactor, $C_{ij} = (-1)^{i+j} M_{ij}$ ($i = k$ or $j = k$); (3) multiply each element in the row or column selected by C_{ij} and sum the results. This is $\det(\mathbf{A})$. Thus, if we selected the kth row,

$$\det(\mathbf{A}) = \sum_{j=1}^{n} a_{kj} C_{kj} = \sum_{j=1}^{n} (-1)^{k+j} a_{kj} M_{kj} \tag{2.25}$$

or the kth column

$$\det(\mathbf{A}) = \sum_{i=1}^{n} a_{ik} C_{ik} = \sum_{i=1}^{n} (-1)^{i+k} a_{ik} M_{ik}. \tag{2.26}$$

The amount of work involved in computing the determinant of an nth order matrix when n is large is awesome for a high speed computing machine. Consequently, we avoid it if at all possible. For example, if **A** is

25×25 then 25! multiplications are required. On a Cray 1, 2.5×10^{-8} seconds are required per multiplication.† Thus, it would take in excess of 10^{10} years to evaluate $|\mathbf{A}|$.

2.3.5 Matrix Inverse and Simultaneous Equations

The inverse of an $n \times n$ matrix is a square matrix $\mathbf{B}$ satisfying

$$\mathbf{AB} = \mathbf{BA} = \mathbf{I}$$

and is denoted by $\mathbf{B} = \mathbf{A}^{-1}$. Not every square matrix has an inverse. If $|\mathbf{A}| \neq 0$ then $\mathbf{A}^{-1}$ exists and we say $\mathbf{A}$ is *invertible* or *nonsingular*. If, on the other hand $|\mathbf{A}| = 0$ then $\mathbf{A}^{-1}$ will not exist and $\mathbf{A}$ is *singular*. The utility of the inverse is seen when we consider the solution of simultaneous algebraic equations. For example, suppose we wish to solve the following set for x, y, and z:

$$5x - 3y + 2z = 14$$
$$x + y - 4z = -7$$
$$7x - 3z = 1.$$

We can write this problem in terms of a matrix equation $\mathbf{Ax} = \mathbf{b}$ where $\mathbf{A}$, called the *coefficient matrix* is

$$\mathbf{A} = \begin{bmatrix} 5 & -3 & 2 \\ 1 & 1 & -4 \\ 7 & 0 & -3 \end{bmatrix} \quad \mathbf{x} = \begin{bmatrix} x \\ y \\ z \end{bmatrix} \quad \text{and } \mathbf{b} = \begin{bmatrix} 14 \\ -7 \\ 1 \end{bmatrix}.$$

For this problem, $|\mathbf{A}| \neq 0$; thus, $\mathbf{A}$ is nonsingular and the solution vector $\mathbf{x}$ can be found by premultiplying the matrix equation by $\mathbf{A}^{-1}$, i.e.,

$$\mathbf{A}^{-1} \mathbf{Ax} = \mathbf{A}^{-1} \mathbf{b}$$
$$\mathbf{Ix} = \mathbf{A}^{-1} \mathbf{b}$$
$$\mathbf{x} = \mathbf{A}^{-1} \mathbf{b}.$$

As we shall subsequently see, techniques are employed to reduce the reservoir fluid flow equations to systems of simultaneous algebraic equations of the form

$$\begin{aligned} a_{11} x_1 + a_{12} x_2 + \ldots + a_{1n} x_n &= b_1 \\ a_{21} x_1 + a_{22} x_2 + \ldots + a_{2n} x_n &= b_2 \\ \vdots \qquad \vdots \qquad\qquad \vdots \quad &\;\; \vdots \\ a_{n1} x_1 + a_{n2} x_2 + \ldots + a_{nn} x_n &= b_n \end{aligned} \quad (2.27)$$

† Assuming vectorized code and 100% efficiency.

which are most conveniently solved by matrix analysis. Eq. 2.27 can be represented in the compressed matrix form $\mathbf{Ax} = \mathbf{b}$ where

$$\mathbf{A} = \begin{bmatrix} a_{11} & a_{12} & \cdots & a_{1n} \\ a_{21} & a_{22} & \cdots & a_{2n} \\ \vdots & \vdots & & \vdots \\ a_{n1} & a_{n2} & \cdots & a_{nn} \end{bmatrix} \quad \mathbf{x} = \begin{bmatrix} x_1 \\ x_2 \\ \vdots \\ x_n \end{bmatrix} \quad \mathbf{b} = \begin{bmatrix} b_1 \\ b_2 \\ \vdots \\ b_n \end{bmatrix}.$$

2.3.6 Matrix Eigenvalue Problem

Many applications of matrices require a solution to the problem $\mathbf{Ax} = \lambda\mathbf{x}$ where $\mathbf{A}$ is n^{th} order, $\mathbf{x}$ is a nonzero vector and λ is a scalar. We want to find, for a given matrix $\mathbf{A}$, those numbers λ such that a matrix multiplication of a vector $\mathbf{x}$ yields the same thing as the scalar multiplication $\lambda\mathbf{x}$. This is known as the *matrix eigenvalue problem* where λ is an *eigenvalue* and $\mathbf{x}$ is the *eigenvector* associated with λ. Since $\mathbf{Ax} = \lambda\mathbf{x}$, we can also write $(\mathbf{A} - \lambda\mathbf{I})\mathbf{x} = \mathbf{0}$. This corresponds to a homogeneous set of n algebraic equations in n unknowns. Obviously $\mathbf{x} = \mathbf{0}$ is a solution (the trivial one); however, we exclude this possibility since we restrict $\mathbf{x}$ to be nonzero. It can be shown that nontrivial solutions will exist if and only if $|\mathbf{A} - \lambda\mathbf{I}| = 0$. The expansion of this determinant yields a polynomial of degree n called the *characteristic polynomial*, $p(\lambda)$ say, whose roots $\{\lambda_i\}_{i=1}^n$ are the eigenvalues we seek. They may be real or complex numbers. The *spectral radius* of matrix $\mathbf{A}$ is defined by

$$\rho(\mathbf{A}) = \max_{1 \leq i \leq n} |\lambda_i|. \tag{2.28}$$

It plays an important role in determining whether or not a stable, convergent solution is possible for a given matrix problem.

2.4 Solution of Simultaneous Linear Algebraic Equations[3,4]

Techniques for solving systems of linear algebraic equations in a reservoir simulator can be broadly categorized as *direct* or *iterative* methods. Each has its particular advantages and disadvantages which will be discussed more fully later.

2.4.1 Gaussian Elimination

We consider solving Eq. 2.27 for the unknowns $x_1, x_2, \ldots, x_n$ assuming the a_{ij}'s and b_i's are known. The method of Gaussian elimination, a direct method, consists of reducing the system of n equations in n unknowns to a system of $(n-1)$ equations in $(n-1)$ unknowns. Next the system of $(n-1)$ equations in $(n-1)$ unknowns is reduced to a system of $(n-2)$ equations in $(n-2)$ unknowns. This process is continued until one obtains one equation in one unknown. Thus, the one unknown is determined. The remaining are found by *back-substitution*.

To illustrate, consider the following example:

$$\begin{align} x_1 + x_2 + x_3 &= 2 \\ 2x_1 + x_2 + x_3 &= 3 \\ -2x_1 + x_2 &= 0. \end{align} \qquad (2.29)$$

Subtract twice the first row from the second row and add twice the first row to the third; thus,

$$\begin{align} x_1 + x_2 + x_3 &= 2 \\ -x_2 - x_3 &= -1 \\ +3x_2 + 2x_3 &= 4. \end{align} \qquad (2.30)$$

Multiply the second row by 3 and add the result to the third row; then the equations reduce to

$$\begin{align} x_1 + x_2 + x_3 &= 2 \\ x_2 + x_3 &= 1 \\ x_3 &= -1. \end{align} \qquad (2.31)$$

Therefore,

$$x_3 = -1.$$

Put the value of $x_3 = -1$ in

$$\begin{align} x_2 + x_3 &= 1 \\ \therefore x_2 &= 2. \end{align}$$

Substitute the values of x_2 and x_3 in $x_1 + x_2 + x_3 = 2$ and one gets $x_1 = 1$.

Note that in performing this procedure, the matrix and right-hand side in Eq. 2.29 were

$$\mathbf{A} = \begin{bmatrix} 1 & 1 & 1 \\ 2 & 1 & 1 \\ -2 & 1 & 0 \end{bmatrix}; \quad \mathbf{b} = \begin{bmatrix} 2 \\ 3 \\ 0 \end{bmatrix}.$$

Prior to performing the back-substitution these were transformed to Eq. 2.31

$$U = \begin{bmatrix} 1 & 1 & 1 \\ 0 & 1 & 1 \\ 0 & 0 & 1 \end{bmatrix}; \quad \underset{\sim}{b} = \begin{bmatrix} 2 \\ 1 \\ -1 \end{bmatrix};$$

i.e., the matrix **A** was converted to an upper triangular matrix. Thus, solution by Gaussian elimination is a triangularization of **A** to yield an upper triangular matrix **U** followed by a back solution for the vector **x**. To achieve this, all elements below the main diagonal of **A** are eliminated.

Some variations of this procedure can be employed. Gauss-Jordan reduction avoids the step of back-substitution. We illustrate this by the same example. The first step is the same as in Gaussian elimination to arrive at Eq. 2.30. We then proceed as follows: multiply the second row of Eq. 2.30 by 3 and 1 and add the results to the third and the first row, respectively

$$\begin{aligned} x_1 + 0x_2 + 0x_3 &= 1 \\ 0x_1 - x_2 - x_3 &= -1 \\ 0x_1 + 0x_2 - x_3 &= 1. \end{aligned} \tag{2.32}$$

Next, add (-1) times the third row to the second row, hence

$$\begin{aligned} x_1 + 0x_2 + 0x_3 &= 1 \\ 0x_1 - x_2 + 0x_3 &= -2 \\ 0x_1 + 0x_2 - x_3 &= 1. \end{aligned} \tag{2.33}$$

Then

$$\begin{aligned} x_3 &= 1/-1 = -1, \\ x_2 &= -2/-1 = 2, \end{aligned}$$

and

$$x_1 = 1/1 = 1.$$

Another technique is to factor matrix **A** into lower and upper triangular matrices **L** and **U**. Thus we have for **Ax** = **b**, **LUx** = **b** where **A** = **LU**. Let **Ux** ≡ **y**, then the problem is solved in the following sequence of steps:

(1) Factorization, ie., find $\mathbf{L} = [l_{ij}]$ and $\mathbf{U} = [u_{ij}]$ where $l_{ij} = 0$, $i < j$ and $u_{ij} = 0$, $i > j$.
(2) Solve **Ly** = **b** for **y**.
(3) Solve **Ux** = **y**.

In (1) **U** will be unit upper triangular, i.e., with ones on the main diagonal. Step (2) is a forward solution for **y** and (3) is a backward solution for **x**. This technique is called **LU** decomposition and is identical with Choleski's (or Banachiewicz's) method for symmetric matrices. We illustrate the procedure for the problem treated before.

The **LU** factors of **A** are

$$\mathbf{L} = \begin{bmatrix} 1 & 0 & 0 \\ 2 & -1 & 0 \\ -2 & 3 & -1 \end{bmatrix}; \mathbf{U} = \begin{bmatrix} 1 & 1 & 1 \\ 0 & 1 & 1 \\ 0 & 0 & 1 \end{bmatrix}.$$

The forward solution involves

$$\begin{bmatrix} 1 & 0 & 0 \\ 2 & -1 & 0 \\ -2 & 3 & -1 \end{bmatrix} \begin{bmatrix} y_1 \\ y_2 \\ y_3 \end{bmatrix} = \begin{bmatrix} 2 \\ 3 \\ 0 \end{bmatrix}.$$

Thus,

$$y_1 = 2$$
$$2y_1 - y_2 = 3 \quad \text{or} \quad y_2 = 1$$
$$-2y_1 + 3y_2 - y_3 = 0, \quad \text{or} \quad y_3 = -1$$

and the back solution is

$$\begin{bmatrix} 1 & 1 & 1 \\ 0 & 1 & 1 \\ 0 & 0 & 1 \end{bmatrix} \begin{bmatrix} x_1 \\ x_2 \\ x_3 \end{bmatrix} = \begin{bmatrix} 2 \\ 1 \\ -1 \end{bmatrix}.$$

Therefore,

$$x_3 = -1, \; x_2 = 2, \text{ and } x_1 = 1.$$

A number of solution techniques employed in reservoir simulators, both direct and iterative, have their basis in the **LU**-decomposition concept. The algorithm for finding the **L** and **U** factors is given in Appendix A.6.

2.4.2 Iterative Methods

We consider finding a solution to the matrix equation $\mathbf{Ax} = \mathbf{b}$ (where **A** is $n \times n$) by iteration. Suppose we divide each row of **A** by its diagonal element (assuming $a_{ii} \neq 0$ for every i); then

$$\mathbf{D\,Ax} = (\mathbf{I} - \mathbf{B})\mathbf{x} = \mathbf{Db} \equiv \mathbf{c} \qquad (2.34)$$

where **D** is a diagonal matrix with $d_{ii} = 1/a_{ii}$, $i = 1, 2, \ldots, n$ and **B** is an $n \times n$ matrix consisting of zeros on the diagonal and the off-diagonals are $-a_{ij}/a_{ii}$, $i \neq j$, $i = 1, 2, 3, \ldots, n$. Thus we can write

$$\mathbf{x} = \mathbf{Bx} + \mathbf{c}. \tag{2.35}$$

A method of successive approximations is given by

$$\mathbf{x}^{(l+1)} = \mathbf{Bx}^{(l)} + \mathbf{c} \tag{2.36}$$

where l is an iteration level ($\mathbf{x}^{(0)}$ is arbitrary). Eq. 2.36 defines a convergent process if for any given $\mathbf{x}^{(0)}$, the sequence $\{\mathbf{x}^{(l)} | l = 1, 2, 3, \ldots\}$ converges. If the spectral radius of B is less than one, then convergence is guaranteed for most iterative processes.

We illustrate the procedure with the example previously employed, i.e., Eq. 2.29. To assure that the diagonals are all nonzero, we rewrite the problem as

$$\begin{aligned} x_3 + x_2 + x_1 &= 2 \\ x_3 + x_2 + 2x_1 &= 3 \\ x_2 - 2x_1 &= 0. \end{aligned} \tag{2.37}$$

Matrix-wise this amounts to a column interchange of the first and last columns. We could have alternatively interchanged the first and last rows. The iteration matrix, **B**, is

$$\mathbf{B} = \begin{bmatrix} 0 & -1 & -1 \\ -1 & 0 & -2 \\ 0 & +\tfrac{1}{2} & 0 \end{bmatrix}$$

and

$$\mathbf{c} = \begin{bmatrix} 1 & 0 & 0 \\ 0 & 1 & 0 \\ 0 & 0 & -\tfrac{1}{2} \end{bmatrix} \begin{bmatrix} 2 \\ 3 \\ 0 \end{bmatrix} = \begin{bmatrix} 2 \\ 3 \\ 0 \end{bmatrix}.$$

If we take as a first guess, $\mathbf{x}^{(0)} = \begin{bmatrix} 1 \\ 1 \\ 1 \end{bmatrix}$ then we get after 10 iterations,

$$\mathbf{x}^{(10)} = \begin{bmatrix} -0.875 \\ 1.75 \\ 0.9375 \end{bmatrix}.$$

Obviously, the rate of convergence is quite slow. However, there are methods for accelerating the convergence rate. This will be discussed later when we treat solution techniques for reservoir models.

2.5 Linear Algebra[5]

Linear algebra provides an overall framework of rules within which we can manipulate vectors, matrices, etc. In the following, we briefly touch on some basic definitions.

2.5.1 Definitions

Let V_n be a vector space consisting of the set of vectors $\{x_i | i = 1, 2, \ldots, n\}$. If there exists a set of scalars $\{\alpha_i | i = 1, 2, \ldots, n\}$ and we form the sum $\mathbf{u} = \sum_{i=1}^{n} \alpha_i x_i$ where $\mathbf{u}$ is also a vector in V_n, then we say that $\mathbf{u}$ is a *linear combination* of the x_i's.

A set of vectors $\{X_i\}_{i=1}^{n}$ is called *linearly independent* if $\sum_{i=1}^{n} \alpha_i x_i = 0$ implies $\alpha_i = 0$ for every i. If the set is not linearly independent then it is linearly dependent, i.e., if there exists some scalars α_i, not all zero such that

$$\sum_{i=1}^{n} \alpha_i x_i = 0, \text{ with some } \alpha_i \neq 0.$$

If every vector in V_n can be written as a linear combination of the set $\{x_i\}_{i=1}^{n}$, i.e., if $\mathbf{u} = \sum_{i=1}^{n} \alpha_i x_i$ for every $\mathbf{u}$ in v_n then we say that the set $\{x_i\}_{i=1}^{n}$ spans the space. For example, the set of vectors **i**, **j**, and **k** spans 3-space. Furthermore, if the set $\{x_i\}_{i=1}^{n}$ is also a linearly independent set, then we say that they form a basis of V_n. A set of vectors may span a vector space and still not be a linearly independent set. However, if the spanning set is a dependent set, we can always extract an independent set that also spans the space. In other words, every spanning set of vectors contains a basis. Once we find a basis of a vector space, we can uniquely represent any other vector in that space as a linear combination of the basis vectors.

2.6 Exercises

1. If $b = 2i + 2j - k$ and α is a scalar number, for what values of α is $|\alpha b| = 1$?
2. Find a unit vector having the same direction as b where $b = \cos \alpha\, i + \sin \alpha\, j$.
3. If $|a| = |b|$ is it necessarily true that $a = b$?
4. If b is nonzero and $\alpha = |a|/|b|$ what can you say about $|\alpha b|$?
5. Find $\nabla \Phi$ if
    (a) $\Phi = xyz$
    (b) $\Phi = e^{xyz}$
    (c) $\Phi = \sin x \cos y$
6. If r and θ are polar coordinates in the xy-plane, determine grad r and grad θ.
7. Find the divergence of the following:
    (a) $xyz(i + j + k)$
    (b) $yz^2\, i - zx^2\, k$
8. Why is the following not valid?
$$(a \cdot b) \cdot c = a \cdot (b \cdot c)$$
9. Expand $(A + B)^2$ if A and B are matrices comformable for addition and multiplication.
10. What must be true about a, b, c, and d if the matrices
$$\begin{bmatrix} a & b \\ c & d \end{bmatrix} \text{ and } \begin{bmatrix} 1 & 1 \\ -1 & 1 \end{bmatrix}$$
are to commute?
11. Denote by E_j an n-vector with 1 in the jth row, all other components being zero. Interpret each of the products below where A is an arbitrary matrix of order n.
    (a) $E_j^T A$ (b) $A E_j$ (c) $E_j^T A E_j$ (d) $E_j^T E_k$ (e) $E_j E_k^T$
12. Show that every nonsymmetric matrix can be written as the sum of a symmetric matrix and a skew-symmetric matrix.
13. Let $A = \begin{bmatrix} \cos \theta & \sin \theta \\ \sin \theta & -\cos \theta \end{bmatrix}$. Find A^2.
14. Given that
$$\begin{bmatrix} 1 & 0 & 0 \\ 0 & -1 & 2 \\ 0 & 0 & -3 \end{bmatrix} \cdot A \cdot \begin{bmatrix} 1 & 0 & 0 \\ 0 & 0 & 1 \\ 0 & 1 & 0 \end{bmatrix} = \begin{bmatrix} 1 & 2 & 3 \\ 4 & 3 & 4 \\ 3 & 2 & 1 \end{bmatrix},$$
find A.
15. If $A = \begin{bmatrix} 3 & 2 & 1 \\ 1 & 0 & 2 \end{bmatrix}$; $B = \begin{bmatrix} -\tfrac{1}{5} & 0 & \tfrac{1}{5} \\ 1 & 0 & 1 \\ \tfrac{2}{5} & 0 & -\tfrac{2}{5} \end{bmatrix}$,

find AB. What is unusual about this result?

16. Find the eigenvalues of the following matrices.

   (a) $A = \begin{bmatrix} 1 & 2 \\ 4 & 3 \end{bmatrix}$ (b) $A = \begin{bmatrix} 1 & -2 \\ 1 & 1 \end{bmatrix}$

   (c) $A = \begin{bmatrix} t & 2t \\ 2t & -t \end{bmatrix}$ (d) $A = \begin{bmatrix} 2 & -1 & 1 \\ 3 & -2 & 1 \\ 0 & 0 & 1 \end{bmatrix}$

17. Is $\begin{bmatrix} 1 \\ 1 \end{bmatrix}$ an eigenvector of $A = \begin{bmatrix} 1 & 2 \\ 3 & 4 \end{bmatrix}$?

18. (a) What is the spectral radius of

$$A = \begin{bmatrix} 0.53 & 0.43 & -0.01 \\ 0 & -0.25 & 0.92 \\ 0 & 0 & .834 \end{bmatrix}$$

   (b) Is A singular or nonsingular?

19. Solve the following problems by Gaussian elimination rounding all calculations to three decimal places. Note the effects of round-off errors by substituting your answers back into the equations. In which problem are the round-off errors worse? Why? What can be done to minimize them?

   (a) $8x_1 - x_2 + x_3 = 9$
   $-x_1 + 15x_2 - x_3 = 26$
   $2x_1 - x_2 + 12x_3 = 36$

   (b) $x_1 + 7x_2 - 9x_3 = -12$
   $6x_1 - x_2 + 15x_3 = 49$
   $-13x_1 + 18x_2 - x_3 = 20$

20. Solve 19(a) using an iterative scheme with $x^{(0)} = [1, 1, 1]^T$. What is your answer after five iterations?

21. Determine if the following are linearly independent or linearly dependent.

   (a) $\left\{ \begin{bmatrix} 2 \\ 6 \\ -2 \end{bmatrix} \begin{bmatrix} 3 \\ 1 \\ 2 \end{bmatrix} \begin{bmatrix} 8 \\ 16 \\ -3 \end{bmatrix} \right\}$

   (b) $\left\{ \begin{bmatrix} 1 \\ 2 \\ 1 \end{bmatrix} \begin{bmatrix} 1 \\ 0 \\ 2 \end{bmatrix} \begin{bmatrix} 1 \\ 1 \\ 0 \end{bmatrix} \right\}$

22. Can the vector $u = \begin{bmatrix} 2 \\ 1 \\ 5 \end{bmatrix}$ be written as a linear combination of those in problem 21?

23. Do the vectors in problem 21 span V_3?

24. Let P_2 be the vector space consisting of all polynomials of degree ≤ 2 and the zero polynomial. Let $X_1 = x^2 + 2x + 1$ and $X_2 = x^2 + 2$. Does $\{X_1, X_2\}$ span P_2?

25. Consider the set $S = \{x^2 + 1, x - 1, 2x + 2\}$. Is it a basis of P_2?

26. Show that every vector in V_n can be uniquely represented as a linear combination of a basis for V_n.

27. Prove that a set of nonzero vectors is linearly dependent if and only if one of the vectors is a linear combination of the others.

2.7 References

1. Lass, H.: *Vector and Tensor Analysis*, McGraw-Hill Book Co. Inc., New York City (1950).
2. Bronson, R.: *Matrix Methods—An Introduction*, Academic Press, New York City (1969).
3. Faddeeva, V.N.: *The Computational Methods of Linear Algebra*, Dover Publications, Inc., New York City (1959).
4. Varga, R.S.: *Matrix Iterative Analysis*, Prentice-Hall, Inc., Englewood Cliffs (1962).
5. Kolman, B.: *Elementary Linear Algebra*, Macmillan, Inc., New York City (1970).

3
Properties of Reservoir Rocks and Fluids

What is your substance, where of are you made,
That millions of strange shadows on you tend?
Since every one, hath every one, one shade,
And you but one, can every shadow lend.

Shakespeare

3.1 Introduction

Reservoir rocks constitute a porous environment within which hydrocarbons and water are found. In this book, attention is confined to three fluid phases: water, oil, and gas which occupy the pore spaces of the rock. Each fluid phase is regarded as a separate continuum wherein certain *macroscopic* properties at points within the continuum can be defined. The term *macroscopic* is used in association with an arbitrarily small reference volume, v_o, at a point P in a reservoir R. Any property assigned to P such as density, viscosity, etc., is considered to be at the geometrical centroid of v_o. All lines passing through any collection of centroids are supposed continuous. Consequently, each fluid phase becomes a fictitiously smooth medium. This level of treatment as opposed to the *microscopic* or *molecular* level is necessary to avoid erratic variations in some of the rock and fluid properties.[1,2] Moreover, the assumption of a separate continuum for each fluid phase restricts us to the treatment of immiscible fluids separated by interfaces across which pressure discontinuities exist.

3.2 Basic Rock and Rock-Fluid Properties

3.2.1 Permeability, Porosity, Saturation and Compressibility

A fundamental property of a porous medium is its ability to conduct fluids. This ability, termed *permeability* and denoted by k, is an empirically determined parameter.[3] *Porosity*, ϕ, at a point P is the ratio of the void volume to the total bulk volume there. It can be expressed as a fraction or a percentage. Clearly, for this definition to be meaningful, the characteristic length across v_o, ℓ say, should satisfy the relationship, $d_\phi \ll \ell \ll L$ where d_ϕ is the average pore diameter and L is a characteristic length over which ϕ may be expected to vary significantly.[4] The *saturation*, S_p, of phase p is the volume fraction of the total void volume at any point P. Thus we have the constitutive relationship at all points P in R,

$$S_w + S_o + S_g = 1. \tag{3.1}$$

The compressibility of the rock can be expressed in terms of the bulk volume at a point.[5] However, in reservoir simulators we express it in terms of the porosity, i.e.,

$$c_r = \frac{1}{\phi}\frac{\partial \phi}{\partial p}. \tag{3.2}$$

If c_r is a constant and ϕ is a function of the reservoir pressure only, then integration of Eq. 3.2 yields

$$\phi = \phi_o e^{c_r(p-p_o)}. \tag{3.3}$$

The first two terms of the Taylor's series expansion of Eq. 3.3 are sufficient to characterize the dependence of porosity on pressure since c_r is a small number (on the order of 10^{-5}/psi). Consequently,

$$\phi \approx \phi_o \{1 + c_r(p - p_o)\}. \tag{3.4}$$

The subscript, o, in Eqs. 3.3 and 3.4 refers to a reference or datum condition. In reservoir situations, $p \gg p_o$, and p_o is frequently neglected. (This is done in chapter 7.) Furthermore, pressure, p, is frequently identified with pressure in a particular fluid phase in multiphase flow systems.

3.2.2 Wettability

When two immiscible fluids contact a solid surface, one of them will tend to spread or adhere to it more so than the other. This is a result of the surface energies between the fluids and the solid. A measure of wettability is the contact angle, θ_c, which is related to the surface energies by the Young-Dupre[6] equation. For example, for a water-oil-solid system, we have

$$\sigma_{os} - \sigma_{ws} = \sigma_{ow} \cos \theta_c \tag{3.5}$$

where

σ_{os} = interfacial energy, oil, and solid.
σ_{ws} = interfacial energy, water, and solid.
σ_{ow} = interfacial energy, oil, and water.
θ_c = contact angle measured into the water, degrees.

Now, $0 \leq \theta_c \leq 180°$. If $\theta_c < 90°$ we say the system is water-wet, if $\theta_c > 90°$ it is oil-wet, and if $\theta_c = 90°$ it is neutral. These situations are depicted in Fig. 3.1.

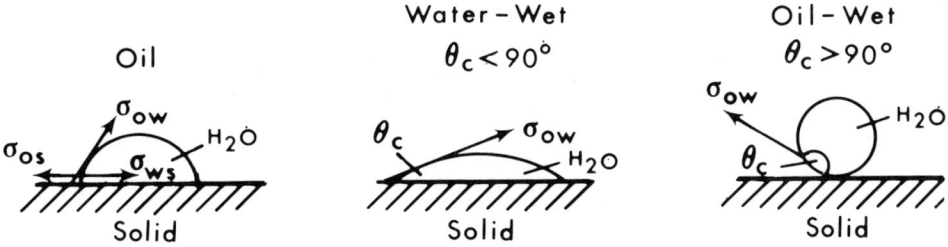

Fig. 3.1 Wettability Configurations.

Wettability of mineral surfaces can be altered by absorbed monolayers of surface-active components or thicker layers of deposited organic materials.[7] Furthermore, a hysteresis effect can occur where the contact angle is not the same when a liquid is advancing as when it is receding on a solid.[8] Thus, a fluid may preferentially wet the solid when advancing but not when it recedes.

3.2.3 Capillary Pressure

Capillary pressure is the pressure difference existing across the interface of two immiscible fluids in a capillary (porous) system. If we take it to be positive, then it is the nonwetting phase pressure minus the wetting phase pressure, i.e., $P_c = p_{nw} - p_w$. Thus, for a water-oil system with water the wetting phase we have

$$P_{cwo} = p_o - p_w \qquad (3.6)$$

and for a gas-oil system,

$$P_{cgo} = p_g - p_o. \qquad (3.7)$$

Laboratory experiments show that capillary pressure can be represented as a single-valued function of one of the phase saturations, e.g., $P_{cwo} = P_{cwo}(S_w)$ and $P_{cgo} = P_{cgo}(S_g)$. Typical dependence of capillary pressure on saturation is shown in Fig. 3.2 for drainage of water (receding) from a porous medium and imbibition (advancing) of the water.

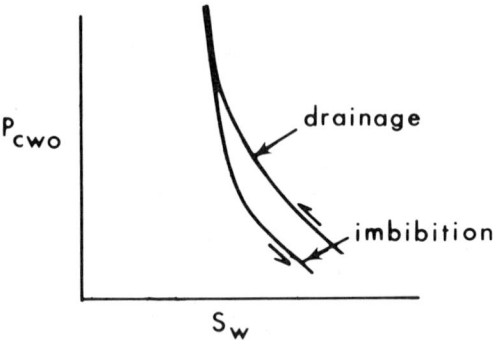

Fig. 3.2 Drainage and Imbibition Capillary Pressure Curves.

The *capillary hysteresis* effect we observe is related to the differences in the advancing and receding contact angles.† Furthermore, during drainage, the larger pores are the first to empty whereas the smaller ones do so reluctantly, if at all. This *capillary retention* explains why the capillary pressure corresponds to a higher saturation on the drainage curve. On imbibition, the smaller pores eagerly fill first, whereas, the larger ones are least apt to, leading to a lower capillary pressure curve.

3.2.4 Relative Permeability

When more than one fluid flows through the porous network in a rock, we encounter relative permeability phenomena. Relative permeability is

---

† The relationship of capillary pressure to the contact angle in porous media is frequently expressed by means of a dimensionless correlating parameter called the Leverett J-function.[9-11]

the ratio of the effective permeability of a given fluid at a fixed saturation to the permeability at 100 percent saturation. By effective permeability we mean the ability of the porous material to conduct a fluid when its saturation is less than 100 percent of the pore space. We write

$$k_{rl} = \frac{k_l}{k} \qquad (3.8)$$

where k_{rl} and k_l are the relative and effective permeabilities of phase l and k is the absolute permeability. The bounds on relative permeability are $0 \leq k_{rl} \leq 1$. Like capillary pressure, laboratory experiments can be performed to determine its dependence on phase saturation. For example, for a water-oil system, k_{ro} and k_{rw} as functions of S_w typically have the appearance shown in Fig. 3.3.

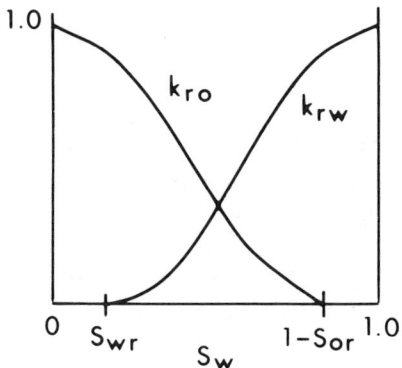

Fig. 3.3 Water-Oil Relative Permeability.

Similar curves are obtained for gas-oil systems. In both cases hysteresis effects are observed depending upon the saturation history. The end points on the relative permeability curves, i.e., the saturations where $k_{rl} \rightarrow 0$ define *irreducible* values. In Fig. 3.3, S_{wr} and S_{or} are the irreducible saturations to water and oil, respectively.

If water, oil, and gas are flowing simultaneously at a point then one requires three-phase relative permeabilities. Laboratory determination of dynamic three-phase relative permeabilities is extremely difficult although some success has been achieved with steady-state methods.[12] In reservoir simulators, two-phase relative permeabilities are used almost exclusively. When needed, three-phase values are computed by the simulator using one of several techniques in the published literature.[13-15] These techniques are based upon the following assumptions:

(1) The water relative permeability is a function of water saturation only regardless of the relative proportions of the gas and oil phases.

(2) The gas relative permeability is a function of gas saturation only regardless of the relative proportions of the water and oil phases.
(3) The oil relative permeability is a function of both the gas and water saturations.

Consequently, k_{rw} and k_{rg} are simply determined from the two-phase water-oil and gas-oil relative permeability data while k_{ro} is computed by an explicit formula. A modified form of Stone's[14] formula is[16]

$$k_{ro} = k^*_{row}\{(k_{row}/k^*_{row} + k_{rw})(k_{rog}/k_{row} + k_{rg}) - (k_{rw} + k_{rg})\}. \qquad (3.9)$$

Here k_{row} and k_{rog} are the values of k_{ro} obtained from the water-oil and gas-oil relative permeability curves, respectively, and $k^*_{row} \equiv k_{row}(S_{wr})$.

3.3 Reservoir Fluid Properties

3.3.1 Characteristics of Black-Oil Reservoir Fluids

If oil can be characterized as a two-component fluid, the reservoir in which it is found is called a *black-oil reservoir*. The two components, in this case, are a dead oil, and the gas dissolved in the oil at reservoir pressure. If R_S is the volume of gas dissolved in a unit volume of oil at a given pressure then the density of the oil is given by

$$\rho_o = (R_s \rho_{gs} + \rho_{os})/B_o \qquad (3.10)$$

where ρ_{gs} and ρ_{os} are the gas and oil densities at surface conditions. We write similar expressions for the water and gas densities,

$$\rho_w = \rho_{ws}/B_w \qquad (3.11)$$

$$\rho_g = \rho_{gs}/B_g \qquad (3.12)$$

B_o, B_w, and B_g are dimensionless quantities reflecting the volumes that a unit volume of oil, water, or gas, respectively, occupy under reservoir conditions. These are called *formation volume factors*.[17] Usually B_o and $B_w \geq 1$ because dissolution of gas into these two phases causes them to swell with increasing pressure. On the other hand, $B_g \leq 1$ because the gas is compressed as pressure increases. The gas formation volume factor can be related, by way of the gas laws, to the reservoir temperature and pressure, i.e.,

$$B_g = (z p_s T)/p T_s \qquad (3.13)$$

where the subscript s refers to standard conditions and z is a gas deviation factor.[17]

The formation volume factors are also related to the fluid compressibility. For example, if c_l is the compressibility for fluid l ($l = w, o, g$), then

$$c_l = -\frac{1}{B_l}\frac{dB_l}{dp}. \qquad (3.14)$$

For oil, the term $(\rho_g B_g/\rho_o B_o)dR_s/dp$ should be added to the above expression for a reservoir pressure, $p < p_s$, the saturation pressure. The latter is the minimum pressure where all the available gas dissolves in the oil. Above the saturation pressure, this term vanishes because R_s becomes a constant. When $p < p_s$ we say the oil is *saturated,* and this state is characterized by a free gas saturation. When $p \geq p_s$, $S_g = 0$, the oil is *undersaturated.* For a saturated condition, the first term in the expression for c_o (i.e., Eq. 3.14) is dominant and $c_o < 0$. This negative compressibility reflects the oil swelling as pressure increases until p_s is reached. For $p > p_s$, $c_o > 0$ since further pressure increases serve only to reduce the volume of the oil. From Eq. 3.14 it is obvious that the reversal in c_o is due to a change in slope, dB_o/dp. Consequently, for $p < p_s$, $dB_o/dp > 0$, and for $p > p_s$, $dB_o/dp < 0$. The same things can be said about the water compressibility and formation volume factor; however, dissolution of gas into water is usually not as significant as that in oil. Orders of magnitude for c_o and c_w are 10^{-5}/psi and 10^{-6}/psi. Values of c_g are roughly equivalent to the reciprocal of pressure and are always positive.

3.3.2 Some Characteristics of Compositional Fluids

In many instances it is not sufficient to characterize an oil as a two-component fluid. For example, if the oil is volatile such that upon a pressure reduction it reverts to a vapor phase, then a multicomponent treatment is required. That is, the concentrations of individual hydrocarbon components (methane, ethane, propane, etc.) partitioning between the vapor and liquid phases must be determined. This is also true for so-called condensate reservoirs.[18] Hydrocarbons that go from liquid to vapor, and vice versa under normal reservoir producing conditions are referred to as *compositional* fluids. In reality, all reservoirs contain compositional fluids, including black-oil systems. However, compositional phenomena in the latter are not sufficiently dominant to justify more costly multicomponent treatment.

Consider one mole of compositional fluid that separates into V moles of vapor and L moles of liquid at a given temperature and pressure. Thus, $L + V = 1$. The mole fraction of a component i in the liquid and vapor phases, x_i and y_i, respectively are related to each other by an equilibrium K-value,

$$K_i = y_i/x_i. \qquad (3.15)$$

Charts of K_i have been compiled as functions of pressure and temperature for a number of different hydrocarbon components.[19] We have the following relationship for an N-component system ($N > 1$):

$$\sum_{i=1}^{N} x_i = \sum_{i=1}^{N} y_i = 1 \qquad (3.16)$$

$$Lx_i + Vy_i = z_i \qquad (3.17)$$

where z_i is the overall mole fraction of the ith component. Assuming a fixed reservoir temperature, when $p = p_s$, then $L = 1$, $V = 0$, and

$$\sum_{i=1}^{N} z_i K_i = 1. \qquad (3.18)$$

Similarly, a *dew point pressure*, p_d, is defined as that pressure where $L = 0$, $V = 1$, and

$$\sum_{i=1}^{N} z_i/K_i = 1. \qquad (3.19)$$

Usually a trial and error procedure is required to compute p_s and p_d using Eqs. 3.18 and 3.19. Burcik[17] gives a treatment of this procedure.

3.4 Exercises

1. Porosity, ϕ, is defined by
$$\phi = \frac{V_p}{V_b} = 1 - \frac{V_s}{V_b}$$
where V_b, V_p and V_s are bulk, pore, and solid (grain) volumes, respectively. The compressibilities are
$$c_b = -\frac{1}{V_b}\frac{\partial V_b}{\partial p}, \quad c_p = -\frac{1}{V_p}\frac{\partial V_p}{\partial p}, \quad \text{and} \quad c_s = -\frac{1}{V_s}\frac{\partial V_s}{\partial p}.$$
Show that the definition of porosity implies $c_b = (1 - \phi)c_s + \phi c_p$.

2. Establish the basis for Eq. 3.2.

3. The reference volume, v_o, also called the *representative elementary volume* (rev), within which one can define properties at a point P in a reservoir R, has a characteristic radius, r_o.
    (a) How would you expect porosity to behave for $r < r_o$ where r approaches zero?

(b) Suppose $r > r_o$ and is increasing gradually. Describe how the porosity may vary as a function of r in homogeneous media. In heterogeneous media.
(c) Let $\hat{v}_o$ be a reference volume in R having a characteristic radius $\hat{r}_o$ in which one can define density of a fluid. If r_o is the minimum radius of an rev for the definition of porosity to be meaningful, then which situation is most likely:
    (1) $\hat{r}_o > r_o$
    (2) $\hat{r}_o = r_o$
    (3) $\hat{r}_o < r_o$
Give a reason for your answer.

4. If you were simulating a waterflood project and had both imbibition and drainage P_c- and k_r-data available, which would you use in the simulator?
5. Suppose you are supplied B_o data as a function of pressure, p, in tabular form. Consider two consecutive values in the table, k and $k + 1$ say, where $(B_o)_k < (B_o)_{k+1}$. If p_k inadvertently is greater than p_{k+1} for p_k and $p_{k+1} < p_s$ what effect can this have on the computation of c_o? What if $p_k < p_{k+1}$ and both are greater than p_s? Are either of these situations physically possible?

3.5 References

1. Corey, A.T.: *Mechanics of Heterogeneous Fluids in Porous Media*, Water Resources Publications, Fort Collins (1977).
2. Bear, J.: *Dynamics of Fluids in Porous Media*, Elsevier North-Holland, Inc., New York City (1972).
3. Amyx, J.W., Bass, D.M., Jr., and Whiting, R.L.: *Petroleum Reservoir Engineering*, McGraw-Hill Book Co. Inc., New York City (1960).
4. Whitaker, S.: "Advances in the Theory of Fluid Motion in Porous Media," *Ind. and Eng. Chem.* (Dec. 1969) **61**, No. 12, 14–28.
5. Collins, R.E.: *Flow of Fluids Through Porous Materials*, Van Nostrand Reinhold Company, New York City (1961).
6. Adams, N.K., *The Physics and Chemistry of Surfaces*, Oxford University Press, London (1941).
7. Mungan, N.: "Interfacial Effects in Immiscible Liquid—Liquid Displacement in Porous Media," *Trans.*, AIME (1966) **237**, 247.
8. Bartell, F.E. and Osterhof, H.J.: "Determination of the Wettability of a Solid by a Liquid," *Ind. and Eng. Chem.* (Nov. 1927) **19**, No. 11, 1277.
9. Leverett, M.C.: "Capillary Behavior in Porous Solids," *Trans.*, AIME (1941) **142**, 159.
10. Rose, W.R. and Bruce, W.A.: "Evaluation of Capillary Character in Petroleum Reservoir Rock," *Trans.*, AIME (1949) **186**, 67.
11. Brown, H.W.: "Capillary Pressure Investigations," *Trans.*, AIME (1951) **192**, 67.
12. Leverett, M.C. and Lewis, W.B.: "Steady State Flow of Gas-Oil-Water Mixtures Through Unconsolidated Sands," *Trans.*, AIME (1941) **142**, 107.
13. Stone, H.L.: "Probability for Estimating Three-Phase Relative Permeability," *Trans.*, AIME (1970) **249**, 214.
14. Stone, H.L.: "Estimation of Three-Phase Relative Permeability and Residual Oil Data," *J. Can. Pet. Tech.* (Oct.–Dec. 1973) 53.
15. Dietrich, J.K. and Bondor, P.L.: "Three Phase Oil Relative Permeability Models," paper SPE 6044 presented at the SPE 51st Annual Technical Conference and Exhibition, New Orleans, Oct. 3–6, 1976.

16. Aziz, K. and Settari, A.: *Petroleum Reservoir Simulation,* Applied Science Publishers, London (1979).
17. Burcik, E.J.: *Properties of Petroleum Reservoir Fluids,* IHRDC, Boston (1979).
18. Craft, B.C. and Hawkins, M.F.: *Applied Petroleum Reservoir Engineering,* Prentice-Hall, Inc., Englewood Cliffs (1959).
19. "Engineering Data Book," Natural Gas Process Supplier's Assoc., ninth edition, Tulsa (1972).

4
Reservoir Flow Equations

There is a pleasure sure
In being mad which none but madmen know.
                                John Dryden

Though this be madness, yet there is method in 't.
                                Shakespeare

4.1 Introduction

To arrive at the basic equations that describe reservoir flow, we make use of the continuity equation (Eq. 2.22), an expression for the superficial flow velocity in a porous medium (Darcy's Law), a mathematical expression for flow potential, and appropriate equations of state. In so doing, we take an Eulerian point of view; i.e., we focus our attention on fixed points of space within the field of flow, in contradistinction to the Lagrangian method, where the coordinates of a moving particle are represented as functions of time.

We furthermore invoke the basic assumptions enumerated below:

(1) Flow is laminar and viscous.
(2) Flow is isothermal.
(3) Electrokinetic effects are negligible.
(4) Diffusion effects are negligible.
(5) Flow is irrotational.

In keeping with assumption (4) we confine our attention to immiscible fluids throughout this book. For the time being, we also restrict ourselves to single-phase flow.

4.2 Flow Potential

To derive an expression for flow potential of a fluid at a point P, we follow Hubbert[1] and define it as the amount of mechanical energy to transform a unit mass from some reference level to an arbitrary level, d. Bear[2] points out that Hubbert's potential from the mathematical viewpoint is really a pseudo-potential. However, we follow common usage and refer to Φ as a potential whether the medium is homogeneous and/or isotropic or not.

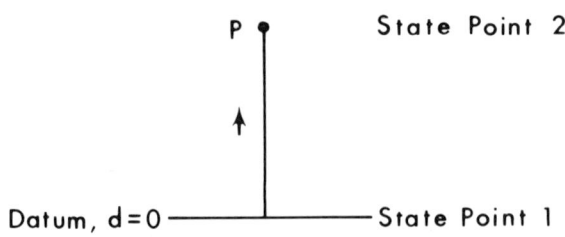

Fig. 4.1   Translation of a Unit Mass.

Consider one unit of mass of fluid translated from state point 1 to state point 2. (Fig. 4.1) The total work done is that incurred by withdrawing the fluid from point 1, lifting it to point 2, compressing it, and finally, rejecting it at point 2. Thus we have

w_1 = work of collection = $-p_0 u_0$ where $u_0 \equiv$ specific volume at $d = 0$.
w_2 = potential energy = $(g/g_c)d$ where g is the gravitational acceleration and g_c is its value at sea-level (32.2 ft/sec^2 in the English system of units).
w_3 = work of compression and rejection
$= -\int_1^2 p\, du + pu$ where $u \equiv$ specific volume at P and the limits on the integral refer to state points.

If, after rejection, the fluid comes to rest (while still compressed), the kinetic energy will be zero. The flow potential is defined by

$$\Phi_h = \sum_{i=1}^{3} w_i$$

or

$$\Phi_h = -p_0 u_0 + \left(\frac{g}{g_c}\right) d - \int_1^2 p\, du + pu.$$

This is a form of Bernoulli's equation for a unit mass. Integrating $\int_1^2 p\,du$ by parts we have

$$\int_1^2 p\,du = [pu]_1^2 - \int_1^2 u\,dp$$

$$\int_1^2 p\,du = pu - p_0 u_0 - \int_1^2 u\,dp.$$

Substituting in the above equation we get

$$\Phi_h = \int_{p_0}^{p} u\,dp + (g/g_c)d.$$

The state references are replaced with pressures on the integral. If we replace u with $1/\rho$, then the flow potential per unit mass of fluid at some point P in R relative to an arbitrary datum is

$$\Phi_h = \int_{p_0}^{p} \frac{1}{\rho(p)}\,dp + (g/g_c)d. \tag{4.1}$$

Eq. 4.1 is Hubbert's potential which is valid for both compressible and incompressible fluids.[1] For either case, the gradient of the potential is

$$\nabla \Phi_h = 1/\rho\, \nabla p + (g/g_c)\, \nabla d. \tag{4.2}$$

Some authors define a potential function, Φ, where $\nabla \Phi = \rho \nabla \Phi_h$ in which case Eq. 4.2 becomes

$$\nabla \Phi = \nabla p + \rho(g/g_c)\nabla d = \nabla p + \gamma \nabla d \tag{4.3}$$

where γ is expressed as pressure per unit distance. If state point 2 is selected as positive downward (as is frequently done in reservoir applications), then the last term on the right-hand sides of Eqs. 4.1–4.3 is preceded by a minus sign.

4.3 Darcy's Law

A basic law for fluid mechanics is Newton's Law for viscous fluids. Consider a fluid flowing along a solid interface as shown in Fig. 4.2 with velocity, v.

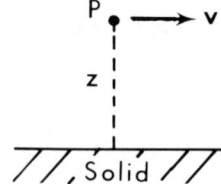

Fig. 4.2 Viscous Flow at P in R.

Newton's Law states

$$d\mathbf{F} = \mu \left(\frac{dv}{dz}\right)_{\text{surface}}$$

where μ = viscosity and $d\mathbf{F}$ is the differential of the viscous force. Since flow in the reservoir is very tortuous, $(dv/dz)_{\text{surface}}$ is very difficult to evaluate and estimations must be used. If we neglect inertial effects, then v is proportional to the flux divided by the bulk external area, A, of the porous element under consideration, i.e.,

$$\mathbf{v} \propto \mathbf{q}/A; \text{ similarly, } \frac{dv}{dz} \propto \mathbf{q}/A.$$

Moreover, if we could integrate over the porous surface, the result would be proportional to the bulk volume, AL, of the porous element. This leads to the conclusion that the total viscous force $\mathbf{F}_\mu$, is given by

$$\mathbf{F}_\mu = \xi\mu \frac{\mathbf{q}}{A} AL$$

where ξ is a proportionality parameter. Thus, $\mathbf{F}_\mu = \xi\mu\rho v L$.

Now consider an element of surface in the neighborhood of the point P as shown in Fig. 4.3.
The component of the resisting force is the viscous force along the axis of the outward drawn normal, i.e.,

$$F_\mu = -\xi\mu\rho \, \mathbf{v} \cdot \mathbf{n} \, dsdz.$$

We use a negative sign because the force is directed opposite to the outward drawn normal. The component of the driving force is a result of a potential gradient at P acting on an element of surface ϕds, i.e.,

$$F_d = \phi \nabla \Phi_h \cdot \mathbf{n} \, dsdz.$$

4.7 Single-Phase Compressible Flow

In this case, the flow equation, after combining Eqs. 4.9–4.11, becomes

$$\nabla \cdot \left\{ \frac{[k]\rho}{\mu} (\nabla p + \gamma \nabla d) \right\} = \frac{\partial}{\partial t}(\phi \rho). \qquad (4.20)$$

If $c = 1/\rho\, \partial \rho/\partial p$, this implies that $\rho \nabla p = 1/c\, \nabla \rho$ so Eq. 4.20 can be written

$$\nabla \cdot \left\{ \frac{[k]}{\mu c} (\nabla \rho + \rho c \gamma \nabla d) \right\} = \phi \frac{\partial \rho}{\partial t} \qquad (4.21)$$

for a system with constant porosity. In cartesian coordinates, we have

$$\frac{\partial}{\partial x}\left(\frac{k_x}{\mu c}\frac{\partial \rho}{\partial x}\right) + \frac{\partial}{\partial y}\left(\frac{k_y}{\mu c}\frac{\partial \rho}{\partial x}\right) + \frac{\partial}{\partial z}\left[\frac{k_z}{\mu c}\left(\frac{\partial \rho}{\partial z} + \rho c \gamma\right)\right] = \phi \frac{\partial \rho}{\partial t}. \qquad (4.22)$$

If again, the medium is homogeneous and isotropic and μ is constant, we get

$$\nabla^2 \rho = \frac{1}{\alpha}\frac{\partial \rho}{\partial t}, \quad \alpha \equiv \frac{k}{\phi \mu c}, \qquad (4.23)$$

neglecting gravity effects. This is known as Fourier's equation or the diffusivity equation. If the fluid is slightly compressible, then one can employ Eq. 4.6 to obtain an equation identical to Eq. 4.23 where ρ is replaced by p. In Eq. 4.23 the constant, α, is frequently called the diffusivity coefficient. Eqs. 4.15–4.23 are generally applicable to single-phase liquid flow (oil or water) in a reservoir.

In treating gas flow, we consider two cases, one an ideal gas, the other a real gas. In both instances, assume the porosity is constant and neglect the source/sink term. Furthermore, the gravity term in Eq. 4.20 is usually negligible; consequently, the flow equation is

$$\nabla \cdot \left(\frac{[k]\rho}{\mu}\nabla p\right) = \phi \frac{\partial \rho}{\partial t} \qquad (4.24)$$

with ϕ constant.

4.7.1 Ideal Gas Flow

For ideal gases, $\rho = pM/RT$, which implies that $\partial \rho/\partial t = (\beta/p)\, \partial p^2/\partial t$ and $\rho \nabla p = \beta \nabla p^2$ where $\beta \equiv M/2\, RT$. Making these substitutions in Eq. 4.24 yields

$$\nabla \cdot \left(\frac{[k]}{\mu} \nabla p^2 \right) = \frac{\phi}{p} \frac{\partial p^2}{\partial t}. \qquad (4.25)$$

From the kinetic theory, the viscosity of an ideal gas is constant in an isothermal environment. As a consequence, we have for isotropic, homogeneous reservoirs

$$\nabla^2 p^2 = \frac{\phi \mu}{kp} \frac{\partial p^2}{\partial t} \qquad (4.26)$$

which, unlike Eq. 4.23 for the liquid flow case, is nonlinear. (See Appendix A.5.1 for the definition of linearity.)

4.7.2 Real Gas Flow

In this case the equation of state is $\rho = pM/zRT$ where z is the gas deviation factor (not to be confused with the cartesian coordinate direction). We employ a Leibenzon transformation.[3] (Al-Hussainy and Ramey,[4] refer to this as the *real gas pseudo pressure function*.)

$$m(p) = \int_{p_0}^{p} \frac{2\lambda \, d\lambda}{\mu(\lambda) z(\lambda)}. \qquad (4.27)$$

Consequently,

$$\frac{\partial m}{\partial t} = \frac{2p}{\mu(p) z(p)} \frac{\partial p}{\partial t}, \qquad (4.28)$$

$$\nabla m = \frac{2p}{\mu(p) z(p)} \nabla p \qquad (4.29)$$

where Leibniz' rule (see Appendix A.4) was applied to Eq. 4.27 to get Eqs. 4.28 and 4.29. Now

$$c = \frac{1}{\rho} \frac{\partial \rho}{\partial p}$$

$$= \frac{zRT}{pM} \frac{\partial}{\partial p} \left(\frac{pM}{zRT} \right)$$

$$= \frac{z}{p} \frac{\partial}{\partial p} \left(\frac{p}{z} \right)$$

$$= \frac{1}{p} - \frac{1}{z} \frac{dz}{dp}. \qquad (4.30)$$

Inserting this equation of state in Eq. 4.24 leads to

$$\nabla \cdot \left(\frac{[k]p}{\mu z} \nabla p\right) = \phi \frac{\partial}{\partial t}\left(\frac{p}{z}\right) \quad (4.31)$$

where the right-hand side can be expanded as follows:

$$\phi \frac{\partial}{\partial t}\left(\frac{p}{z}\right) = \phi \left(\frac{1}{z}\frac{\partial p}{\partial t} - \frac{p}{z^2}\frac{dz}{dp}\frac{\partial p}{\partial t}\right)$$

$$= \phi \frac{p}{z}\left(\frac{1}{p} - \frac{1}{z}\frac{dz}{dp}\right)\frac{\partial p}{\partial t}$$

$$= \phi \frac{p}{z} c \frac{\partial p}{\partial t}$$

$$= \frac{\phi \mu c}{2} \frac{\partial m}{\partial t}. \quad (4.32)$$

To arrive at Eq. 4.32, we used Eqs. 4.28 and 4.30. Similarly, we use Eq. 4.29 to substitute for $p\nabla p/\mu z$ in the left-hand side of Eq. 4.31 and get finally

$$\nabla \cdot ([k] \nabla m) = (\phi \mu c)_i \frac{\partial m}{\partial t}. \quad (4.33)$$

The subscript of the coefficient on the right-hand side indicates this quantity is to be evaluated at initial conditions. For ideal media $[k] = k$, a constant, and then

$$\nabla^2 m = \frac{(\phi \mu c)_i}{k} \frac{\partial m}{\partial t}. \quad (4.34)$$

Eq. 4.34 resembles Fourier's equation, Eq. 4.23; however, it is nonlinear if the coefficient on the right-hand side is not constant.

4.8 Multiphase Flow—The Generalized Flow Equation

We generalize the principles outlined thus far to construct appropriate equations for multiphase flow. To do so, consider flow of a single component,

i, at a point P in R which conceivably is found in all three fluid phases, i.e., in water, oil, and gas. We will have occasion to use indices, 1, 2, and 3 to represent the water, oil, and gas, respectively.

Let C_{ig}, C_{io}, C_{iw} be the mass fractions of component i, in the three phases, then

$$\sum_{l=1}^{3} C_{il}\, \rho_l\, v_l = \text{the mass flux density at point } P \text{ of component } i.$$

$$\phi \sum_{l=1}^{3} C_{il}\, \rho_l\, S_l = \text{the mass of component } i \text{ per unit pore volume.}$$

The continuity equation then becomes

$$\nabla \cdot \left(\sum_{l=1}^{3} C_{il}\, \rho_l v_l \right) \pm \sum_{l=1}^{3} C_{il}\, \tilde{g}_l = -\frac{\partial}{\partial t}\left(\phi \sum_{l=1}^{3} C_{il}\, \rho_l\, S_l \right). \quad (4.35)$$

From a modification of Darcy's Law for multiphase flow we have

$$v_l = -\frac{[k]k_{rl}}{\mu_l}\, \rho_l\, \nabla \Phi_l,\ l = 1, 2, 3. \quad (4.36)$$

The modifier is the relative permeability, k_{rl}, where $0 \le k_{rl} \le 1$. The Hubbert flow potential is (suppressing subscript, h)

$$\Phi_l = \int_{p_0}^{p_l} \frac{d\lambda}{\rho_l(\lambda)} - (g/g_c)d \quad (4.37)$$

where the negative sign on the gravity term is used to denote that the positive direction is downward. Combining Eqs. 4.35–4.37 gives us the generalized flow equation for component i in a three-phase environment, i.e.,

$$\nabla \cdot \left\{ \sum_{l=1}^{3} C_{il}\, \rho_l\, \frac{[k]k_{rl}}{\mu_l}(\nabla p_l - \gamma_l \nabla d) \right\} \pm \sum_{l=1}^{3} C_{il}\, \tilde{g}_l$$

$$= \frac{\partial}{\partial t}\left(\phi \sum_{l=1}^{3} C_{il}\, \rho_l\, S_l \right). \quad (4.38)$$

If there are N components at point P in region R, then we have $3N + 15$ unknowns. These are as follows:

Reservoir Flow Equations

Unknown	Number
C_{il}	$3N$
ρ_l	3
k_{rl}	3
μ_l	3
p_l	3
S_l	3
	Total = $3N + 15$

To obtain a well-determined system, we require $3N + 15$ auxiliary relations. These are obtained from the following:

Equations	Number
$\sum_{i=1}^{N} C_{il} = 1.0,\ l = 1, 2, 3$	3
$\rho_l = \rho(T, p_l)$	3
$k_{rl} = k_r(S_o, S_w, S_g)_l$	3
$\mu_l = \mu(T, p_l)$	3
$P_{cwo} = p_o - p_w$	1
$P_{cgo} = p_g - p_o$	1
$\sum_{l=1}^{3} S_l = 1.0$	1
$\dfrac{C_{ig}}{C_{io}} = K_{igo}(T, p_g, p_o, C_{ig}, C_{io})$	N
$\dfrac{C_{ig}}{C_{iw}} = K_{igw}(T, p_g, p_w, C_{iw}, C_{ig})$	N
Mass balances	N
	Total = $3N + 15$

The symbols K represent the equilibrium constants between the vapor and liquid phases.

In most cases, the hydrocarbon phase changes involving water are minimal and can be ignored. Thus, the system represented by Eq. 4.38 is frequently reduced to one involving fewer equations and unknowns. This is done by considering one "component" to exist in the water phase, i.e., all of the water phase is the *water component*. If there are N hydrocarbon components partitioning exclusively between the gas and oil then the total number of components is $N + 1$, the $(N + 1)^{st}$ being water. Thus, in Eq. 4.38, we let i take on values $1, 2, \ldots, N + 1$ where when the $(N + 1)^{st}$ value is reached, we replace subscript i with w. To restrict hydrocarbon phase transfers with water, set $C_{iw} = 0$ for $i \leq N$, in Eq. 4.38.† Furthermore, $C_{(N+1),1} = C_{ww} = 1.0$. The mass fractions of the hydrocarbon phases are expressed in terms of mole fractions by dividing each term in Eq. 4.38

---

† Alternatively, one achieves the same effect by defining $K_{igw} = 0,\ i \leq N$.

by the appropriate molecular weights M_i, and recognizing that $C_{io}/M_i = x_i n_t$ and $C_{ig}/M_i = y_i n_t$, where n_t is the total number of moles. When this is done the system of equations becomes

$$\nabla \cdot \left\{ \rho_o x_i \frac{[k]k_{ro}}{\mu_o}(\nabla p_o - \gamma_o \nabla d) + \rho_g y_i \frac{[k]k_{rg}}{\mu_g}(\nabla p_g - \gamma_g \nabla d) \right\} \pm (x_i \tilde{g}_o + y_i \tilde{g}_g)$$
$$= \frac{\partial}{\partial t}\{\phi(x_i \rho_o S_o + y_i \rho_g S_g)\}, \quad i = 1, 2, \ldots, N \quad (4.39)$$

$$\nabla \cdot \left\{ \rho_w \frac{[k]k_{rw}}{\mu_w}(\nabla p_w - \gamma_w \nabla d) \right\} \pm \tilde{g}_w = \frac{\partial}{\partial t}(\phi \rho_w S_w) \quad (4.40)$$

Recall the x_i's and y_i's are the mole fractions in the liquid and vapor phases, respectively (see Chapter 3). Eqs. 4.39 and 4.40 form the basis of a *compositional simulator.*[5]

4.9 Black-Oil Simulator

Here we convert Eq. 4.38 to a volumetric material balance. Again we consider one component to exist in the water phase. Similarly, one component exists in the gas phase, namely the gas. Thus, $C_{ww} = C_{gg} = 1.0$. We assume the oil phase consists of only two components, the dissolved gas and the residual or black oil that remains when this gas is liberated. We also invoke the following assumptions:

(1) No phase transfers occur between the water and oil.
(2) No phase transfers occur between the water and gas.
(3) A one-way phase transfer occurs between the gas and oil, i.e., gas moves in and out of the oil, but the oil does not vaporize into the gas phase. Thus we exclude the possibility of gas condensate or volatile oil systems.†

These assumptions lead to the following:

$$C_{ww} = 1 \qquad C_{ow} = 0 \qquad C_{gw} = 0$$

$$C_{wo} = 0 \qquad C_{oo} = \frac{m_o}{m_o + m_g} \qquad C_{go} = \frac{m_g}{m_o + m_g}$$

$$C_{wg} = 0 \qquad C_{og} = 0 \qquad C_{gg} = 1 \qquad (4.41)$$

† In some instances, a black-oil model can be used to approximate compositional systems.[6] However, these applications are limited.

Here m_o, m_g are the masses of oil and gas, respectively, and V_o, V_g are their respective volumes. Subscript s denotes a surface condition. R_s and B_o, B_w, etc., are the gas solubility in oil and formation volume factors, respectively. Now $B_o = V_o/V_{os} = V_o\rho_{os}/m_o$, but $V_o = (m_o + m_g)/\rho_o$ so $B_o = (m_o + m_g)\rho_{os}/m_o\rho_o$, and $C_{oo} = m_o/(m_o + m_g)$. Therefore,

$$C_{oo} = \frac{\rho_{os}}{\rho_o B_o}. \tag{4.42}$$

Moreover, $R_s = V_{gs}/V_{os} = (m_g/\rho_{gs})/(m_o/\rho_{os}) = m_g\rho_{os}/m_o\rho_{gs}$. Now $C_{go} = [m_o/(m_o + m_g)][m_g/m_o] = C_{oo}V_g\rho_{gs}/V_o\rho_{os} = C_{oo}R_s\rho_{gs}/\rho_{os}$. Consequently,

$$C_{go} = \frac{R_s\rho_{gs}}{\rho_o B_o}. \tag{4.43}$$

Also $B_w = \rho_{ws}/\rho_w$ and $B_g = \rho_{gs}/\rho_g$. If we use these relationships and those in Eqs. 4.41–4.43 in Eq. 4.38 and expand the equations, we obtain the following:

Water Equation:

$$\nabla \cdot \left\{ \frac{[k]k_{rw}}{\mu_w B_w} (\nabla p_w - \gamma_w \nabla d) \right\} \pm Q_w = \phi \frac{\partial}{\partial t}\left(\frac{S_w}{B_w}\right) \tag{4.44}$$

Oil Equation:

$$\nabla \cdot \left\{ \frac{[k]k_{ro}}{\mu_o B_o} (\nabla p_o - \gamma_o \nabla d) \right\} \pm Q_o = \phi \frac{\partial}{\partial t}\left(\frac{S_o}{B_o}\right) \tag{4.45}$$

Gas Equation:

$$\nabla \cdot \left\{ \frac{[k]k_{ro}R_s}{\mu_o B_o} (\nabla p_o - \gamma_o \nabla d) \right\} + \nabla \cdot \left\{ \frac{[k]k_{rg}}{\mu_g B_g} (\nabla p_g - \gamma_g \nabla d) \right\}$$
$$\pm (R_s Q_o + Q_g) = \phi \frac{\partial}{\partial t}\left(\frac{S_g}{B_g} + \frac{S_o R_s}{B_o}\right) \tag{4.46}$$

In Eqs. 4.44–4.46, $Q_l \equiv \tilde{g}_l/\rho_l B_l$, $l = w, o, g$. A computer program that generates approximate solutions to Eqs. 4.44–4.46 is a black-oil reservoir simulator.

4.10 Exercises

1. Suppose a chemical is injected into a reservoir where chemical reaction within the rock is predominant. What effect would this have on the concept of flow potential? Is the concept of flow potential as given in this chapter adequate for thermal recovery processes?
2. Discuss the importance of orienting a reservoir grid such that it coincides with the principal axes of permeability.
3. Show that if $\tilde{g} = 0$ and $\phi = 1$, Eq. 4.9 can also be written as

$$\frac{\partial \rho}{\partial t} + \rho \nabla \cdot \mathbf{v} + \nabla \rho \cdot \mathbf{v}.$$

    Hint: Remember ∇ is both a vector and a differential operator.
4. From the relationship, $c = -1/V \, \partial V/\partial p$:
   (a) Show that $c = 1/\rho \, \partial \rho/\partial p$.
   (b) Show that the expression in (a) implies $\rho \nabla p = 1/c \, \nabla \rho$.

    Hint: For part (b), consider a position vector, $\mathbf{r} = x(s)\mathbf{i} + y(s)\mathbf{j} + z(s)\mathbf{k}$ at some point P in R where s is the arc length of some path through P and $\left|\frac{d\mathbf{r}}{ds}\right| = 1$. First show that

$$c = \frac{\nabla \rho \cdot \frac{d\mathbf{r}}{ds}}{\rho \nabla p \cdot \frac{d\mathbf{r}}{ds}}.$$

5. Show that if a porous medium is itself compressible, the equation of continuity for mass flow has the form

$$\text{div}(\rho \mathbf{v}) = -\phi\left(1 - \frac{c_p}{c}\right)\frac{\partial \rho}{\partial t}$$

    where $c_p = -1/\phi \, \partial \phi/\partial p$ is the pore compressibility and c is the fluid compressibility.
6. If the flow potential is given by

$$\Phi = \int_{p_o}^{p} \frac{d\lambda}{\rho^2(\lambda)},$$

    derive the vector form of the single-phase flow equation without sources or sinks. The final equation should be expressed in terms of only one dependent variable.
7. Develop an expression for Darcy's Law in a one-dimensional bed having dip angle, θ, assuming single-phase, laminar, viscous, isothermal, incompressible fluid flow.
8. Show that Eq. 4.22 follows from Eq. 4.20.
9. Develop the differential equation for real gas flow in a 1-D system without sources or sinks assuming the permeability is pressure dependent according to the Klinkenberg equation,

$$k(p) = k_\infty(1 + b/p)$$

    where k_∞ and b are constants.

10. Suppose in a black-oil reservoir substantial amounts of gas dissolve in water as well as the oil. If R_{sw} and R_{so} are the water and oil gas solubilities, determine the form of the flow equations for this system.

4.11 References

1. Hubbert, M.K.: "The Theory of Ground-Water Motion," *The J. of Geol.* (Nov.–Dec. 1940) **68,** No. 8, 785–944.
2. Bear, J.: *Dynamics of Fluids in Porous Media,* Elsevier North-Holland, Inc., New York City (1972).
3. Leibenzon, L.S.: "Subsurface Hydraulics of Water, Oil and Gas," *Publ. Acad. Nauk.* (1953) 2.
4. Al-Hussainy, R. and Ramey, H.J., Jr.: "The Flow of Real Gases Through Porous Media," *Trans.,* AIME (1966) **237,** 637.
5. Kazemi, H., Vestal, C.R., and Shank, G.D.: "An Efficient Multicomponent Numerical Simulator," *Soc. Pet. Eng. J.* (Oct. 1978) 355.
6. Cook, R.E., Jacoby, R.H., and Ramesh, A.B.: "A Beta-Type Reservoir Simulator for Approximating Compositional Effects During Gas Injection," *Soc. Pet. Eng. J.* (Oct. 1974) 471.

5
Finite Difference Approximations

For he, by geometric scale
Could take the size of pots of ale, . . .
And wisely tell what hour o' th' day
The clock doth strike . . .
By algebra.

Samuel Butler

5.1 Introduction

When we specify a coordinate system (as determined by the nature of the reservoir problem) for flow equations such as those in Eqs. 4.44–4.46, we obtain sets of partial differential equations coupled through auxiliary relationships to achieve a well-determined system. Generally these equations are too complex to be solved analytically. Indeed, only in the idealized cases of single-phase flow can analytic solutions be obtained. As a consequence, one must resort to approximate methods of solution. The most common approach is to apply finite difference techniques. Briefly, this involves superimposing a grid on the region of interest, expressing the partial derivatives in terms of algebraic approximations, and solving the resulting set of algebraic equations. If u represents the exact solution, and v the approximate solution, then one hopes that by making the grid spacings sufficiently small, v will be a satisfactory approximation to u at every grid point.

We begin our discussion by first developing finite difference approximations to partial derivatives using Taylor polynomials for functions of class C^k (see Appendix A.2). This enables us to readily determine the local truncation error. Furthermore, this approach also applies to functions that are not analytic or regular, i.e., functions of class C^∞. We follow this by considering a simple single-phase flow problem in a 1-D cartesian coordinate system to illustrate several methods for solving the algebraic problem. Finally, some attention is devoted to the problems of stability and convergence.

5.2 Finite Differences

Suppose we have a function $f(x)$ of class C^k defined on an interval (a,b) where (a,b) belongs to the set S of real numbers. Then we can represent $f(x)$ by its Taylor polynomial,

$$f(x_o + \Delta x) = f(x_o) + \frac{\Delta x}{1!} f'(x_o) + \frac{\Delta x^2}{2!} f''(x_o) + \frac{\Delta x^3}{3!} f^{(3)}(x_o)$$
$$+ \ldots + \frac{\Delta x^k}{k!} f^{(k)}(\xi), \dagger \quad (5.1)$$

where $\Delta x = x - x_o$, $x_o \in (a,b)$, $\xi = x_o + \theta \Delta x$, $0 < \theta < 1$.

If $k = 4$ then

$$f(x_o + \Delta x) = f(x_o) + \frac{\Delta x}{1!} f'(x_o) + \frac{\Delta x^2}{2!} f''(x_o) + \frac{\Delta x^3}{3!} f^{(3)}(x_o) + \frac{\Delta x^4}{4!} f^{(4)}(\xi_1). \quad (5.2)$$

Similarly we have

$$f(x_o - \Delta x) = f(x_o) - \frac{\Delta x}{1!} f'(x_o) + \frac{\Delta x^2}{2!} f''(x_o) - \frac{\Delta x^3}{3!} f^{(3)}(x_o)$$
$$+ \frac{\Delta x^4}{4!} f^{(4)}(\xi_2). \quad (5.3)$$

Adding Eqs. 5.2 and 5.3 and solving for $f''(x_o)$ we get

$$f''(x_o) = \frac{f(x_o - \Delta x) - 2f(x_o) + f(x_o + \Delta x)}{\Delta x^2} - \frac{\Delta x^2}{4!} \{f^{(4)}(\xi_1) + f^{(4)}(\xi_2)\}. \quad (5.4)$$

If we choose Δx sufficiently small, then the second derivative at the point x_o belonging to (a,b) can be approximated by

$$f''(x_o) \approx \frac{f(x_o - \Delta x) - 2f(x_o) + f(x_o + \Delta x)}{\Delta x^2} \quad (5.5)$$

---

† Here and in the sequel, Δx^k means $(\Delta x)^k$, $k = 2, 3, \ldots$

where the error is

$$\frac{\Delta x^2}{4!}\{f^{(4)}(\xi_1)+f^{(4)}(\xi_2)\}.$$

This is called the local truncation error and is of $O(\Delta x^2)$. (See Appendix A.1 where $g(x) \equiv \Delta x^2$ and $f(x)$ is the expression above.)

We emphasize that Eq. 5.5 is an approximation, hence the use of the symbol $\approx$. However, subsequently, we employ the symbol $=$, and recognize the finite difference representation as an approximation. The process by which we arrived at Eq. 5.5 is called *discretization;* i.e., while the left-hand side is a continuous function for all $x_o \in (a,b)$, the right-hand side is evaluated at discrete points in (a,b), namely, $x_o - \Delta x$, x_o and $x_o + \Delta x$ for a given choice of x_o. To simplify the notation, suppress the arguments of f and employ an index i to refer to the point at x_o, $i+1$ for the point ahead, and $i-1$ for the point behind as depicted below:

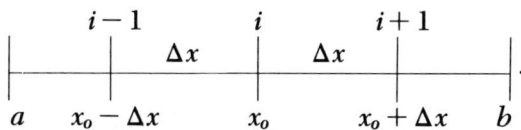

Thus, Eq. 5.5 can be represented as

$$f_i'' = \frac{f_{i-1}-2f_i+f_{i+1}}{\Delta x^2}. \tag{5.6}$$

If the region of interest is 2-D on which we define a function $f(x,y)$, then we employ a double index (i,j) to refer to some point in the region. Thus, the coordinates (x,y) are given by $x = i\Delta x$ and $y = j\Delta y$ for equally spaced grids. We then have

$$(f_{xx})_{ij} = \frac{f_{i-1,j}-2f_{ij}+f_{i+1,j}}{\Delta x^2} \tag{5.7}$$

and

$$(f_{yy})_{ij} = \frac{f_{i,j-1}-2f_{ij}+f_{i,j+1}}{\Delta y^2}. \tag{5.8}$$

Similarly for 3-D regions, we carry a set of triple indices (i,j,k) to refer to a grid point. For discretization of a function dependent on space and time, e.g., $f(x,y,t)$, let $t = n\Delta t$ and write

$$(f_{xx})_{ij}^n = \frac{f_{j+1,j}^n - 2f_{ij}^n + f_{j+1,j}^n}{\Delta x^2}. \tag{5.9}$$

Note, the superscript on f does not represent a power but rather a time-level.

For first order derivatives, we let $k = 2$ in Eq. 5.1 and neglect terms of second order and higher to obtain

$$f_i' = \frac{f_{i+1} - f_i}{\Delta x} \tag{5.10}$$

or alternatively,

$$f_i' = \frac{f_i - f_{i-1}}{\Delta x} \tag{5.11}$$

which are first order correct, i.e., the truncation error is $0(\Delta x)$. We call Eqs. 5.10 and 5.11 the *forward difference* and *backward difference* approximations, respectively. A second order correct approximation can be obtained by letting $k = 3$ in Eq. 5.1 to obtain

$$f(x_o + \Delta x) = f(x_o) + \frac{\Delta x}{1!} f'(x_o) + \frac{\Delta x^2}{2!} f''(x_o) + \frac{\Delta x^3}{3!} f^{(3)}(\xi_1) \tag{5.12}$$

$$f(x_o - \Delta x) = f(x_o) - \frac{\Delta x}{1!} f'(x_o) + \frac{\Delta x^2}{2!} f''(x_o) - \frac{\Delta x^3}{3!} f^{(3)}(\xi_2) \tag{5.13}$$

Subtracting,

$$\frac{f(x_o + \Delta x) - f(x_o - \Delta x)}{2\Delta x} = f'(x_o) + 0(\Delta x)^2.$$

Thus,

$$f_i' = \frac{f_{i+1} - f_{i-1}}{2\Delta x}.$$

This is called a *central difference* approximation.

5.3 Application to Single-Phase Flow

To illustrate finite difference techniques, consider an application to single-phase fluid flow in a homogeneous, isotropic, 1-D cartesian system. Assume the viscosity is constant and the fluid is slightly compressible. We then have

$$\frac{\partial^2 p}{\partial X^2} = \frac{1}{\alpha}\frac{\partial p}{\partial \tau} \quad 0 < X < L, \quad \tau > 0. \tag{5.14}$$

Eq. 5.14 can be normalized by using the transformations, $u = p/p_i$, $x = X/L$, $t = \alpha\tau/L^2$ where p_i is the pressure at $\tau = 0$.
Thus we get

$$u_{xx} = u_t, \quad 0 < x < 1, \quad t > 0 \tag{5.15}$$

on which we impose the auxiliary conditions (after Smith[1]).

$$\left.\begin{array}{l} (1) \ u = 0, \ x = 0, \ t > 0 \\ (2) \ u = 0, \ x = 1, \ t > 0 \\ (3) \ u = 2x, \ 0 \leq x \leq \tfrac{1}{2} \\ = 2(1-x), \ \tfrac{1}{2} \leq x \leq 1 \end{array}\right\} t = 0 \tag{5.16}$$

In Eq. 5.16, (1) and (2) are boundary conditions and (3) is an initial condition. Eq. 5.15 is parabolic (see Appendix A.5) and Eqs. 5.15–5.16 constitute a well-posed problem.[2]

5.3.1 Explicit Method of Solution

Substituting the finite difference approximations for the derivatives in Eq. 5.15 we get

$$\frac{u_{i-1}^n - 2u_i^n + u_{i+1}^n}{(\Delta x)^2} = \frac{u_i^{n+1} - u_i^n}{\Delta t}. \tag{5.17}$$

Now multiply both sides by Δt and define $\gamma = \Delta t/(\Delta x)^2$;

then

$$u_i^{n+1} = \gamma u_{i-1}^n + (1 - 2\gamma) u_i^n + \gamma u_{i+1}^n. \tag{5.18}$$

Eq. 5.18 is an *explicit* formula for u_i at time level $n + 1$. It is expressed in terms of known quantities on the right-hand side. For example, when $n = 0$, the right-hand side is determined by the initial condition for $1 \leq i \leq N - 1$ where N is the number of partitions on the interval $0 \leq x \leq 1$, i.e., $\Delta x = 1/N$. The values of u at $i = 0$ and $i = 1$ are given by the boundary conditions. Notice we have the following computational star:

to propagate the solution from time level n to $n + 1$. Repeated application of the star to all points i on the grid advances the solution to the next time level. The process is again repeated for successive time levels. For example, if $\Delta x = 0.1$, $\Delta t = 0.001$ then $\gamma = 0.1$ and a smoothly evolving solution is obtained. On the other hand, if $\Delta x = 0.1$ and $\Delta t = 0.01$ such that $\gamma = 1.0$, wild oscillations appear after a few time steps which are not dampened with time. This is referred to as the *instability* problem and is obviously dependent on the value of γ. We shall see later that the explicit formulation requires that $0 \leq \gamma \leq \frac{1}{2}$ to achieve a stable, convergent solution.

The truncation error in Eq. 5.18 is $0(\Delta x^2 + \Delta t)$. To see the effects of this, compare the numerical solution to results obtained from the analytical solution given by

$$u(x,t) = \frac{8}{\pi^2} \sum_{k=1}^{\infty} \left(\sin \frac{k\pi}{2} \right) (\sin k\pi x) \frac{e^{-k^2\pi^2 t}}{k^2}. \quad (5.19)$$

The two tables below are for the points $x = 0.3$ and $x = 0.5$, respectively when $\gamma = 0.1$. The error is the difference between the solutions expressed as a percentage of the analytical solution.

TABLE 5.1 $x = 0.3$ (After Smith[1])

t	FDS	AS	Diff.	% Error
0.005	0.5971	0.5966	0.0005	0.08
0.01	0.5822	0.5799	0.0023	0.40
0.02	0.5373	0.5334	0.0039	0.70
0.10	0.2472	0.2444	0.0028	1.10

(FDS) ≡ Finite Difference Solution; AS ≡ Analytical Solution

TABLE 5.2 $x = 0.5$ (After Smith[1])

t	FDS	AS	Diff.	% Error
0.005	0.8597	0.8404	0.0193	2.3
0.01	0.7867	0.7743	0.0124	1.6
0.02	0.6891	0.6809	0.0082	1.2
0.10	0.3056	0.3021	0.0035	1.2

(FDS) ≡ Finite Difference Solution; AS ≡ Analytical Solution

Finite Difference Approximations

The error in the tables is actually a combination of truncation error and *round-off* error. The latter results because the computational procedure is not capable of producing the exact solution to the difference equations. To do so would require the computer to retain an infinite number of digits. The error in Table 5.2 is greatest at early time because the first derivative of the initial condition is discontinuous at $x = 0.5$. On the other hand, at $x = 0.3$, the early time effect is absent since the initial condition is a smooth function there. At later times in both cases, the error appears to stabilize at about 1.2%.

5.3.2 Crank-Nicolson Method[3]

The following difference operators are defined as:

$$\Delta^2 u \equiv u_{i-1} - 2u_i + u_{i+1}$$

$$\Delta_t u \equiv u_i^{n+1} - u_i^n.$$

Thus the explicit formulation can be represented as

$$\gamma \Delta^2 u^n = \Delta_t u. \tag{5.20}$$

Because of the instabilities associated with Eq. 5.20, Crank and Nicolson[3] suggested replacing the left-hand side with an average value between the $(n+1)^{st}$ and n^{th} time levels, i.e.,

$$\gamma(\Delta^2 u^{n+1} + \Delta^2 u^n)/2 = \Delta_t u. \tag{5.21}$$

After expanding and rearranging Eq. 5.21 we get

$$-\gamma u_{i-1}^{n+1} + 2(1+\gamma) u_i^{n+1} - \gamma u_{i+1}^{n+1} = \gamma u_{i-1}^n + 2(1-\gamma) u_i^n + \gamma u_{i+1}^n. \tag{5.22}$$

The computational star for Eq. 5.22 is depicted below:

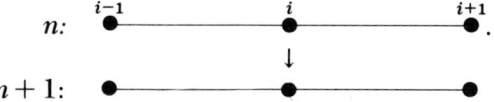

Since a single value of the dependent variable cannot be computed explicitly at $n+1$, this is an *implicit* method. Eq. 5.22 generates a set of $N-1$ algebraic equations that can be expressed in matrix form as

$$\mathbf{Au} = \mathbf{d} \tag{5.23}$$

where

$$A = \begin{bmatrix} b_1 & -c_1 & & & & \\ -a_2 & b_2 & -c_2 & & \bigcirc & \\ & -a_3 & b_3 & -c_3 & & \\ & & \cdot & \cdot & \cdot & \\ & \bigcirc & & \cdot & \cdot & \cdot \\ & & & & -a_{N-1} & b_{N-1} \end{bmatrix} \quad u = \begin{bmatrix} u_1 \\ u_2 \\ u_3 \\ \vdots \\ u_{N-1} \end{bmatrix} \quad d = \begin{bmatrix} d_1 \\ d_2 \\ d_3 \\ \vdots \\ d_{N-1} \end{bmatrix}$$

$b_i = 2(1 + \gamma)$, $a_i = c_i = \gamma$, $i = 1, 2, \ldots, N-1$. The matrix A is a *tridiagonal* matrix. Eq. 5.21 is readily solved by Gaussian elimination using an algorithm, described in section 5.3.3 that excludes operations on zero elements.

5.3.3 Thomas' Algorithm[4]

The procedure involves successively eliminating the unknowns, u_r, say, from the $(r+1)$st equations by appropriate algebraic manipulations (where $r = 1, 2, \ldots, N-2$). In so doing the system is actually transformed to an equivalent problem where the matrix is unit upper triangular as discussed in chapter 2. Such a system is easily solved by back-substitution for u_r, $r = N-2, N-3, \ldots, u_2, u_1$. To illustrate, from the first equation we have

$$u_1 = \frac{d_1 + c_1 u_2}{b_1}.$$

Substituting this in the second and rearranging yields

$$\left\{ b_2 - \frac{a_2 c_1}{b_1} \right\} u_2 - c_2 u_3 = d_2 + a_2 \frac{d_1}{b_1}.$$

At any later stage of the elimination we have the form

$$\alpha_{i-1} u_{i-1} - c_{i-1} u_i = S_{i-1},$$

thus,

$$u_{i-1} = \frac{S_{i-1} + c_{i-1} u_i}{\alpha_{i-1}}. \tag{5.24}$$

Substituting this in the ith equation we get, after rearranging

$$\left\{ b_i - \frac{a_i c_{i-1}}{\alpha_{i-1}} \right\} u_i - c_i u_{i+1} = d_i + \frac{a_i S_{i-1}}{\alpha_{i-1}}$$

where

$$a_i = b_i - \frac{a_i c_{i-1}}{\alpha_{i-1}} \qquad (5.25)$$

$$S_i = d_i + \frac{a_i S_{i-1}}{\alpha_{i-1}} \qquad (5.26)$$

and

$$\left.\begin{array}{l}\alpha_1 = b_1 \\ S_1 = d_1\end{array}\right\} \qquad (5.27)$$

Eqs. 5.25–5.26 are forward recurrence formulas that can be used to generate the α- and S-arrays using Eq. 5.27 for starting values. Once these arrays are obtained, the backward recursion formula, Eq. 5.24, is used to compute u_{i-1}, $i = N$, $N-1$, $N-2$, ..., 1 subject to the constraint, $c_{N-1} = 0$. This technique is known as the Thomas algorithm for tridiagonal matrices.[4] Similar algorithms have been presented by others;[5,6] however, they are not as efficient as Thomas' algorithm, though less round-off error may be incurred.[5] Where applicable, the Thomas algorithm is used almost exclusively in reservoir simulation problems.

5.4 Stability Analysis

We can examine the stability of a finite difference approximation by one of two methods; (a) a matrix method which requires the eigenvectors of the matrix; (b) a method that relies upon Fourier analysis (the von Neumann method[7]). We consider the matrix method first for the explicit formulation given in Eq. 5.18. Expanding,

$$\begin{array}{l}u_1^{n+1} = \qquad\qquad (1-2\gamma)\,u_1^n \;+ \gamma u_2^n \\ u_2^{n+1} = \gamma u_1^n + (1-2\gamma)\,u_2^n \;+ \gamma u_3^n \\ u_3^{n+1} = \gamma u_2^n + (1-2\gamma)\,u_3^n \;+ \gamma u_4^n \\ \qquad\cdot\qquad\quad\cdot\qquad\qquad\cdot \\ \qquad\cdot\qquad\quad\cdot\qquad\qquad\cdot \\ \qquad\cdot\qquad\quad\cdot\qquad\qquad\cdot \\ u_{N-1}^{n+1} = \gamma u_{N-2}^n + (1-2\gamma)\,u_{N-1}^n\end{array} \qquad (5.28)$$

$$\begin{bmatrix} (1-2\gamma) & \gamma & & & \bigcirc \\ \gamma & (1-2\gamma) & \gamma & & \\ & \cdot & \cdot & \cdot & \\ & & \cdot & \cdot & \cdot \\ \bigcirc & & & \gamma & (1-2\gamma) \end{bmatrix} \begin{bmatrix} u_1^n \\ u_2^n \\ \cdot \\ \cdot \\ u_{N-1}^n \end{bmatrix} = \begin{bmatrix} u_1^{n+1} \\ u_2^{n+1} \\ \cdot \\ \cdot \\ u_{N-1}^{n+1} \end{bmatrix} \quad (5.29)$$

Thus we have

$$\begin{aligned} \mathbf{u}^{n+1} &= \mathbf{A}\,\mathbf{u}^n \\ &= \mathbf{A}(\mathbf{A}\,\mathbf{u}^{n-1}) \\ &= \mathbf{A} \cdot \mathbf{A}(\mathbf{A}\,\mathbf{u}^{n-2}) \\ &= \mathbf{A} \cdot \mathbf{A} \cdot \mathbf{A} \ldots \mathbf{A}\,\mathbf{u}^0, (n+1) \text{ times} \\ &= \mathbf{A}^{n+1}\,\mathbf{u}^0. \end{aligned}$$

In general, we write $\mathbf{u}^n = \mathbf{A}^n \mathbf{u}^0$ where the superscript on $\mathbf{A}$ denotes a power and not time discretization (see Appendix A.6.3). Now suppose at time level zero, an error is introduced such that instead of $\mathbf{u}^0$, the exact initial condition, we use $\mathbf{u}_*^0$. Then $\mathbf{u}_*^n = \mathbf{A}^n \mathbf{u}_*^0$. The error vector at time level n is defined by

$$\begin{aligned} \mathbf{e}^n &= \mathbf{u}^n - \mathbf{u}_*^n \\ &= \mathbf{A}^n \mathbf{u}^0 - \mathbf{A}^n \mathbf{u}_*^0 \\ &= \mathbf{A}^n (\mathbf{u}^0 - \mathbf{u}_*^0) \\ &= \mathbf{A}^n \mathbf{e}^0 \end{aligned}$$

where $\mathbf{e}^0$ is an $N-1$ vector and can be considered an element of an $N-1$ dimensional vector space. Then, $\mathbf{A}$ is a matrix of a linear transformation on that vector space. Consequently, if we can find $N-1$ linearly independent eigenvectors of $\mathbf{A}$ that span the space, they will constitute a basis, and we can express $\mathbf{e}^0$ as a linear combination of them. In view of the theorems given in Appendix A.6.2, such a basis can be found.

Let (v_i) be the basis derived from $\mathbf{A}$ and let (α_i) be a corresponding set of scalars, $i = 1, 2, \ldots, N-1$.

Then,

$$\mathbf{e}^0 = \sum_{i=1}^{N-1} \alpha_i\, v_i$$

or
$$\mathbf{e}^n = \sum_{i=1}^{N-1} \alpha_i \mathbf{A}^n \mathbf{v}_i$$

$$= \sum_{i=1}^{N-1} \alpha_i \lambda_i^n \mathbf{v}_i \qquad (5.30)$$

where λ is an eigenvalue of $\mathbf{A}$. Eq. 5.30 directly follows from the second theorem cited in Appendix A.6.3 where $f(\mathbf{A}) = \mathbf{A}^n$. Clearly, $\mathbf{e}^n$ will be bounded if $\rho(\mathbf{A}) \leq 1$. If $\rho(\mathbf{A}) > 1$ then the system is divergent and unstable. Now $\mathbf{A}$, in Eq. 5.29 can be factored into $\mathbf{A} = \mathbf{I} + \gamma \mathbf{B}$ where

$$\mathbf{B} = \begin{bmatrix} -2 & 1 & & & & \\ 1 & -2 & 1 & & \bigcirc & \\ & 1 & -2 & 1 & & \\ & & \cdot & \cdot & \cdot & \\ & \bigcirc & & \cdot & \cdot & \cdot \\ & & & & 1 & -2 \end{bmatrix}$$

It can be shown that the eigenvalues of $\mathbf{B}$ are

$$-4 \sin^2 \frac{k\pi}{2N}, \quad k = 1, 2, \ldots N-1.$$

The corresponding eigenvectors are

$$\mathbf{v}_k = \left(\sin \frac{k\pi}{N}, \sin \frac{2k\pi}{N}, \ldots, \sin \frac{(N-1)k\pi}{N} \right).$$

Since $\mathbf{A} = \mathbf{I} + \gamma \mathbf{B} \equiv f(\mathbf{B})$ then the eigenvalues of $\mathbf{A}$ are

$$1 - 4\gamma \sin^2 \frac{k\pi}{2N}, \quad k = 1, 2, \ldots N-1.$$

For stability and convergence we require that $\rho(\mathbf{A}) \leq 1$, i.e., $-1 < 1 - 4\gamma \sin^2 \frac{\pi k}{2N} < 1$. The right-hand side of this inequality puts no restrictions on γ. However, from the left-hand side we find that $\gamma \leq \frac{1}{2}$.

Stability analysis by matrix methods cannot be universally applied, especially to the highly nonlinear problems we treat in reservoir simulation. For this reason, frequent use is made of the von Neumann method which is simpler, but less rigorous because it neglects the effects of boundary conditions.[1] This is treated next.

Any function $f(x)$ satisfying Dirichlet's conditions[8] can be represented by

$$f(x) = \sum_{k=0}^{\infty} \left(a_k \cos \frac{k\pi x}{L} + b_k \sin \frac{k\pi x}{L} \right)$$

on an interval (0, L), i.e., a Fourier series. Substituting the Euler expansions for $\sin \theta$ and $\cos \theta$ in the above equation gives[8]

$$f(x) = \sum_{k=-\infty}^{\infty} A_k e^{\frac{Jk\pi x}{L}}, J = \sqrt{-1}. \tag{5.31}$$

An error with finite norm in the $N - 1$ dimensional vector space for a problem of the form $\mathbf{A}\mathbf{u}^n = \mathbf{u}^{n+1}$ can be shown to satisfy Dirichlet's conditions.[7] Let E_i be the spatial error at some point i; thus,

$$E_i = \sum_{k=-\infty}^{\infty} A_k e^{\frac{Jk\pi x}{L}}. \tag{5.32}$$

Furthermore, let $x = i\Delta x$ and $L = N\Delta x$ and define $\beta_k \equiv k\pi/N\Delta x$. Then Eq. 5.32 can be written

$$E_i = \sum_{k=-\infty}^{\infty} A_k \epsilon_i(k) \tag{5.33}$$

where $\epsilon_i(k) \equiv e^{J\beta_k i\Delta x}$. For our purposes, we examine only one component ϵ_i in Eq. 5.33 rather than E_i. The error component of the space-time continuum must be such that it reduces to ϵ_i at $t = 0$. Consequently, we write

$$\epsilon_i^n = e^{J\beta_k i\Delta x} e^{n\alpha\Delta t} \tag{5.34}$$

where α is some number, real or complex. If $\zeta = e^{\alpha\Delta t}$, then the error will be bounded as n increases provided $|\zeta| \leq 1$. This is known as the *von Neumann criterion*.[1]

Again let $\mathbf{u}$ be the exact solution to Eq. 5.20 and $\mathbf{u}_*$ the calculated value, then

$$u_i^n = u_{*i}^n \pm \epsilon_i^n. \tag{5.35}$$

If Eq. 5.35 is substituted into Eq. 5.20 one gets

$$\gamma \Delta^2 u_*^n \pm \gamma \Delta^2 \epsilon_i^n = \Delta_t u_* \pm \Delta_t \epsilon_i^n \tag{5.36}$$

and since $\gamma \Delta^2 u_*^n = \Delta_t u_*$, then

$$\gamma \Delta^2 \epsilon_i^n = \Delta_t \epsilon_i^n, \qquad (5.37)$$

i.e., the error component satisfies the difference equation. Now substitute Eq. 5.34 in Eq. 5.37. After some algebra, we find

$$\gamma(e^{-J\beta_k \Delta x} + e^{J\beta_k \Delta x} - 2) = \zeta - 1. \qquad (5.38)$$

The left-hand side $= -2\gamma [1 - \cos(\beta_k \Delta x)]$

$$= -4\gamma \sin^2\left(\frac{\beta_k \Delta x}{2}\right),$$

consequently,

$$\zeta = 1 - 4\gamma \sin^2\left(\frac{\beta_k \Delta x}{2}\right). \qquad (5.39)$$

Applying the von Neumann criterion, we find once again that $\gamma \leq \frac{1}{2}$ is required for stability.

5.4.1 Stability Analysis of the Crank-Nicolson Method

For the Crank-Nicolson method, Eq. 5.22 leads to a matrix problem of the form

$$\begin{bmatrix} 2(1+\gamma) & -\gamma & & \\ -\gamma & 2(1+\gamma) & -\gamma & \\ & \cdot & \cdot & \cdot \\ & & \cdot & \cdot & \cdot \\ & & & -\gamma & 2(1+\gamma) \end{bmatrix} \begin{bmatrix} u_1^{n+1} \\ u_2^{n+1} \\ \cdot \\ \cdot \\ u_{N-1}^{n+1} \end{bmatrix} = \begin{bmatrix} 2(1-\gamma) & \gamma & & \\ \gamma & 2(1-\gamma) & \gamma & \\ & \cdot & \cdot & \cdot \\ & & \cdot & \cdot & \gamma \\ & & & \gamma & 2(1-\gamma) \end{bmatrix} \begin{bmatrix} u_1^n \\ u_2^n \\ \cdot \\ \cdot \\ u_{N-1}^n \end{bmatrix}$$

If we define the matrix **B** as before the above reduces to

$$(2\mathbf{I} - \gamma \mathbf{B}) \mathbf{u}^{n+1} = (2\mathbf{I} + \gamma \mathbf{B}) \mathbf{u}^n.$$

Since $|2\mathbf{I} - \gamma \mathbf{B}| \neq 0$ for $\gamma > 0$ then $(2\mathbf{I} - \gamma \mathbf{B})$ is nonsingular and the inverse exists. Premultiplying both sides by $(2\mathbf{I} - \gamma \mathbf{B})^{-1}$ gives

$$\mathbf{u}^{n+1} = (2\mathbf{I} - \gamma \mathbf{B})^{-1} (2\mathbf{I} + \gamma \mathbf{B}) \mathbf{u}^n$$

or $\quad \mathbf{u}^{n+1} = \mathbf{A} \mathbf{u}^n \quad$ where $\mathbf{A} = (2\mathbf{I} - \gamma \mathbf{B})^{-1} (2\mathbf{I} + \gamma \mathbf{B}) = f(\mathbf{B}).$

The eigenvalues of **B** are cited after Eq. 5.30; therefore, the eigenvalues of **A** will be

$$\lambda_k = \frac{2 - 4\gamma \sin^2\left(\frac{k\pi}{2N}\right)}{2 + 4\gamma \sin^2\left(\frac{k\pi}{2N}\right)}, \quad k = 1, 2, \ldots, N-1 \qquad (5.40)$$

and for every $\gamma > 0$, $|\lambda_k| < 1$, $k = 1, 2, \ldots, N-1$. Consequently, the Crank-Nicolson method is unconditionally stable.

5.5 Truncation Error

Truncation or *discretization* error is the departure of a finite difference approximation from the solution of a partial differential equation at a grid point. It is possible to determine the local truncation error assuming the computational procedure is capable of producing exactly the solution of the finite difference equation. In practice, this assumption is never true since a computer has a *fixed word-length*, i.e., the number of digits retained is fixed (usually 7 to 12 significant digits in single precision). Thus, any number with more significant digits than a computer can retain is approximated by a rounded value. This leads to *round-off error*. Usually, some upper bound can be found for the local truncation errors for a particular discretization. On the other hand the analysis of round-off error is extremely complex and their magnitudes are not readily predictable. Here we examine only those errors related to truncation; however, the reader should be aware that round-off errors can sometimes be appreciable. Fortunately, there are steps one can take to minimize the effects of the latter which we touch on later.

If we let $L\{u\}_i^n$ represent the partial differential equation with the derivatives evaluated at $x = i\Delta x$, $t = n\Delta t$, and $L_\Delta\{u\}$ its corresponding finite difference form, then the truncation error, Υ, is defined by

$$\Upsilon = L_\Delta\{u\} - L\{u\}_i^n. \qquad (5.41)$$

We say a finite difference method converges if $\Upsilon \to 0$ as Δx and Δt tend to zero. For example, consider the partial differential equation given in Eq. 5.15. Here $L\{u\}_i^n = (u_{xx} - u_t)_i^n$ and $L_\Delta\{u\} = \Delta^2 u/\Delta x^2 - \Delta_t u/\Delta t$ for the explicit formulation. Appealing to Eqs. 5.2 and 5.3 we have

$$u_{i\pm 1}^n = u_i \pm \Delta x (u_x)_i^n + \frac{\Delta x^2}{2!}(u_{xx})_i^n \pm \frac{\Delta x^3}{3!}(u_{xxx})_i^n + \frac{\Delta x^4}{4!}u_{xxxx}^n(\xi_1) \qquad (5.42)$$

such that

Finite Difference Approximations

$$\Delta^2 u/\Delta x^2 = \left(\frac{\partial^2 u}{\partial x^2}\right)_i^n + \frac{\Delta x^2}{12}\frac{\partial^4}{\partial x^4}[u^n(\xi_1) + u^n(\xi_2)]. \tag{5.43}$$

Also,

$$\Delta_t u/\Delta t = \left(\frac{\partial u}{\partial t}\right)_i^n + \frac{\Delta t}{2}\left[\frac{\partial^2 u(\xi_3)}{\partial t^2}\right]_i \tag{5.44}$$

where $\xi_3 = t_o + \theta\Delta t$, $0 < \theta < 1$. Consequently,

$$\Upsilon = \frac{\Delta x^2}{12}\frac{\partial^4 \bar{u}^n}{\partial x^4} - \frac{\Delta t}{2}\left[\frac{\partial^2 u(\xi_3)}{\partial t^2}\right]_i \tag{5.45}$$

where $\bar{u}^n = u^n(\xi_1) + u^n(\xi_2)$. From Eq. 5.15, $u_{xxxx} = u_{tt}$, thus, Eq. 5.45 is approximately

$$\Upsilon = \left(\frac{\Delta x^2}{12} - \frac{\Delta t}{2}\right)\frac{\partial^2 u}{\partial t^2} = \left(\frac{1}{6} - \gamma\right)\frac{\Delta x^2}{2}\frac{\partial^2 u}{\partial t^2}. \tag{5.46}$$

If we had used sixth and third degree Taylor polynomials in Eqs. 5.42 and 5.44, respectively, then Eq. 5.46 would contain an additive term $0(\Delta t^2 + \Delta x^4)$. Thus, in the special case when $\gamma = 1/6$, the explicit formulation is $0(\Delta t^2 + \Delta x^4)$ correct; otherwise, it is $0(\Delta t + \Delta x^2)$. One can also show[9] that over a time domain $0 \leq t \leq T$, the maximum discretization error is $\leq T|\Upsilon|$ where Υ is given by Eq. 5.46 provided $0 < \gamma \leq \frac{1}{2}$. Consequently, this constraint on γ is a sufficient condition for convergence of the explicit method.

5.6 Other Considerations

In this chapter, we briefly touched on the essentials involved in approximating solutions to partial differential equations by using finite differences. We emphasize that the results obtained are indeed approximations. How good they are is dependent upon the spatial and time increments we employ, and the word length of the computing machine. Furthermore, the way we formulate the finite difference expressions determines whether or not a computational algorithm will be stable or not. In analyzing a given formulation for stability, we most often resort to the von Neumann method because of its simplicity. However, under some circumstances, instabilities can be introduced by the initial or boundary conditions which the von Neumann technique will not detect. Short of employing the matrix technique, which may involve a tedious search for eigenvalues, one can appeal to the theorems of Gerschgorin[10] or Brauer[11] cited in Appendices A.6.4 and A.6.5, which give an upper bound, S_m, on the spectral radius of a matrix. A sufficient condition for a stable process is $S_m \leq 1$. Brauer's theorem provides a tighter upper bound than Gerschgorin's. In either case, the values of S_m are easily and quickly computed.

There are some circumstances where stable, numerical processes converge to the solution of the wrong partial differential equation. Such a process is *inconsistent* or *incompatible*. For example, the Dufort-Frankel scheme given by

$$2[u_{i+1}^n - (u_i^{n+1} + u_i^{n-1}) + u_{i-1}^n] = u_i^{n+1} - u_i^{n-1} \tag{5.47}$$

can be used to approximate Eq. 5.15. However, it can be shown by examining the truncation error that Eq. 5.47 is compatible with Eq. 5.15 only when $\gamma \Delta x \to 0$ as $\Delta x \to 0$. If $\gamma \Delta x \to$ a constant, c say, as $\Delta x \to 0$, then Eq. 5.47 converges to the solution of the telegraph equation,

$$u_{xx} = c^2 u_{tt} + u_t. \tag{5.48}$$

5.7 Exercises

1. The Buckley-Leverett approach to predict the behavior of a waterflood in 1-D involves treating an equation of the form:
$$\frac{\partial f}{\partial x} = -\frac{\partial S}{\partial t}, 0 \leq x \leq L, t > 0$$
where f is the fractional flow of water and is considered as a single-valued function of the water saturation, S.
   (a) Write a finite difference expression for this equation that
      (1) treats f implicitly,
      (2) is second order correct in space, and
      (3) is first order correct in time.
   (b) Let $\gamma \equiv \Delta t / \Delta x$, $\tilde{S}_i \equiv$ the exact solution of the difference equation, $\epsilon_i^n \equiv$ the actual computer solution containing error, ϵ_i^n, i.e., $S_i^n - \epsilon_i^n = \tilde{S}_i^n$. Show that
   $$\gamma \{(f_{i+1}^{n+1} - \tilde{f}_{i+1}^{n+1}) - (f_{i-1}^{n+1} - \tilde{f}_{i-1}^{n+1})\} = -2(\epsilon_i^{n+1} - \epsilon_i^n)$$
   where $\tilde{f} \equiv f(\tilde{S})$.
   (c) Assume that $f \epsilon C^1$ then show that
   $$f_{i+1}^{n+1} = \tilde{f}_{i+1}^{n+1} + \epsilon_{i+1}^{n+1} f'(\bar{S})$$
   where $\bar{S}$ is some saturation between S and $\tilde{S}$.
   (d) Show that the error equation is
   $$\tilde{\gamma}(\epsilon_{i+1}^{n+1} - \epsilon_{i-1}^{n+1}) = -2(\epsilon_i^{n+1} - \epsilon_i^n)$$
   where $\tilde{\gamma} \equiv \gamma f'(\bar{S})$.

(e) Perform a stability analysis of the finite difference scheme specified in part (a) and state whether it is (1) unconditionally unstable, or (2) unconditionally stable, or (3) conditionally stable.
 For this you can treat $f'(\bar{S})$ as a constant that is always greater than zero.
2. Suppose for the equation in problem 1(a) you used a finite difference scheme that
 (1) treats f explicitly,
 (2) is second order correct in space, and
 (3) is second order correct in time.
 Show that the truncation error is given by
$$1/6 \{S_{xxt}(\Delta x)^2 - S_{ttt}(\Delta t)^2\}.$$
3. Apply harmonic analysis to determine the stability of the Crank-Nicolson approximation to $u_{xx} = u_t$.
4. Apply harmonic analysis to the difference equation
$$(u_{i-1}^n - 2u_i^n + u_{i+1}^n)/(\Delta x)^2 = (u_i^{n+1} - u_i^n)/2\Delta t.$$
 Is there any choice of Δt for which stability is assured?
5. Consider a finite difference treatment of the equation
$$u_{xx} - \lambda u_x = u_t, \ 0 \leq x \leq 1, \ t > 0, \ \lambda \text{ constant}.$$
 (a) Discuss the classification (see Appendix A.5) of this equation when (1) λ is a very small real number, and (2) when λ is extremely large.
 (b) When λ is very large, oscillations can occur in the computed results even when an implicit formulation is employed. Based on your answer to (a) what is a possible reason?
 (c) It has been found[12] that these oscillations are controlled if u_x is approximated by
$$u_x = \frac{3u_i - 4u_{i-1} + u_{i-2}}{2\Delta x}$$
 which is second order correct. Develop this approximation from Taylor polynomials.
 (d) Show that discretization of the partial differential equation using this approximation for u_x leads to algebraic equations of the form
$$-a_i u_{i-2} - b_i u_{i-1} + c_i u_i - d_i u_{i+1} = g_i, \ i = 1, 2, \ldots, n.$$
 (e) What is the structure of the matrix generated by these equations?
 (f) Develop an algorithm to solve such matrix problems using Gaussian elimination.
 (g) What boundary condition specifications are required to get the solution started?
 (h) Devise a scheme to combine your algorithm with the Thomas algorithm for the first few grid blocks such that only a single condition is required on the near boundary.
 (i) Develop a computer program to execute your solution procedure and use it to solve the partial differential equation subject to the following conditions:
$$u(x,0) = 0, \ 0 \leq x \leq 1$$
$$u(0,t) = 1, \ t > 0$$
$$u(1,t) = 0, \ t > 0$$
 where $\Delta x = 0.05$ and Δt is arbitrarily selected.
6. (a) Show that when the explicit equation (Eq. 5.18) is used to approximate $u_{xx} = u_t$ and it is assumed u possesses continuous and finite derivatives up

to order three in t and order six in x, then the discretization error is the solution of the difference equation

$$\Upsilon_i^{n+1} = \gamma \Upsilon_{i-1}^n + (1 - 2\gamma)\Upsilon_i^n + \gamma \Upsilon_{i+1}^n + \Delta \xi(x,t)$$

where

$$\xi(x,t) = \frac{\Delta x^2}{\Delta t}\left(6\gamma \frac{\partial^2 u}{\partial t^2} - \frac{\partial^4 u}{\partial x^4}\right)_{i,j} + \frac{\Delta t^2}{6}\frac{\partial^3 u}{\partial t^3}(x_i, t_i + \theta_n \Delta t)$$
$$- \frac{\Delta x^4}{360}\frac{\partial^6 u}{\partial x^6}(x_i + \theta_i \Delta x, t_n), \quad -1 < \theta_i < 1, \ 0 < \theta_n < 1.$$

(b) If the maximum value of $|\xi|$ is M deduce for $0 < \gamma \leq \frac{1}{2}$ that $|\Upsilon_i^n| \leq tM$ and show that the spatial discretization error is $O(\Delta x)^2$ except when $\gamma = \frac{1}{6}$ when it is $O(\Delta x)^4$.

7. (a) Show that both Gerschgorin's and Brauer's theorems (see Appendices A.6.4–A.6.5) establish stability for Eq. 5.18 when $\gamma \leq \frac{1}{2}$ but give no useful result when $\gamma > \frac{1}{2}$.

  (b) Show that Gerschgorin's theorem is inadequate for establishing unconditional stability of the Crank-Nicolson approximation (Eq. 5.22).

5.8 References

1. Smith, G.D.: *Numerical Solution of Partial Differential Equations*, Oxford University Press, London (1965).
2. Garabedian, P.R.: *Partial Differential Equations*, John Wiley & Sons, New York City (1964).
3. Crank, J. and Nicolson, P.: "A Practical Method for Numerical Evaluation of Solutions of Partial Differential Equations of the Heat-Conduction Type," *Proc. Camb. Phil. Soc.* (Jan. 1947) **43**, No. 264, 50–67.
4. Richtmyer, R.D. and Morton, K.W.: *Difference Methods for Initial-Value Problems*, second edition, Interscience Publishers, New York City (1967) **2**.
5. Tang, I.C.: "A Simple Algorithm for Solving Linear Equations of a Certain Type," *Zeitschrift für Angewandte Math. und Mech.* (Aug. 1969) **8**, No. 49, 508.
6. Evans, D.J.: "An Algorithm for the Solution of Symmetric General Three Term Linear Systems," *Compt. J.* (Nov. 1971) **14**, No. 4, 444.
7. O'Brien, G.G., Hyman, M.A., and Kaplan, S.: "A Study of the Numerical Solution of Partial Differential Equations," *J. Math. Phys.* (Jan. 1951) **29**, No. 70, 223.
8. Kreysig, E.: *Advanced Engineering Mathematics*, second edition, John Wiley & Sons, New York City (1967) **2**.
9. Carnahan, B., Luther, H.A., and Wilkes, J.O.: *Applied Numerical Methods*, John Wiley & Sons, New York City (1969).
10. Gerschgorin, S.: "Über die Abrenzung der Eigenwerte einer Matrix," *Izv. Akad. Nauk SSSR* (Ser. Mat. 7, 1931) **16**, 749.
11. Brauer, A.: "Limits for the Characteristic Roots of a Matrix, II," *Duke Math. J.* (1947) **14**, 21.
12. Price, H.S., Varga, R.S., and Warren, J.E.: "Application of Oscillation Matrices to Diffusion—Convection Equations," *J. Math. & Phys.* (Sept. 1966) **45**, No. 3, 301–311.

6
Single-Phase Multidimensional Flow

As yet a child,
Nor yet a fool to fame,
I lisped in numbers,
For the numbers came.

Alexander Pope

6.1 Introduction

In the previous chapter we treated a simple single-phase flow problem in 1-D. Also we considered two solution techniques, the explicit method and an implicit method (Crank-Nicolson) for solving the associated matrix problem. Both these techniques are examples of *direct* solution methods, i.e., no iteration was involved. Because of time-step limitations and the associated stability problem, completely explicit methods are rarely employed in reservoir simulation. More often we resort to implicit or semi-implicit methods, or possibly implicit methods combined with explicit computations. We can also employ iterative procedures as opposed to direct methods. However, the principal iterative techniques are deferred to this chapter where we discuss their application to multidimensional problems.

In a multidimensional environment, the basic approach is the same as that applied to the 1-D problem in chapter 5. That is, we begin with a partial differential equation (or collection of them) and replace it (them) with finite difference representations. This ultimately leads to a matrix problem, $\mathbf{Au} = \mathbf{b}$, to be solved. We must be concerned with the stability of the solution process, and convergence if the solution technique is iterative. In most cases, both are assured when the spectral radius of $\mathbf{A}$ or an associated matrix is bounded above by unity.

Before discussing specific single-phase flow problems, we first give a brief description of the grid systems most frequently employed in multidimensional problems. This is followed by a treatment of various alternating direction implicit procedures for formulating the algebraic problem. These require nothing more than a tridiagonal algorithm to achieve a solution. Attention is then turned toward the banded matrix problems which are typically encountered in reservoir simulation. Direct solution of these is discussed including a brief treatment of sparse matrix techniques. Finally, a discussion of two iterative methods, the strongly implicit procedure and successive overrelaxation is presented. Although the presentation is primarily in the context of a particular 2-D parabolic problem, these methods are equally applicable to any parabolic or elliptic problem in both 2-D and 3-D.

6.2 Grid Systems and Boundary Conditions

As we have already indicated, one can relate the finite difference equations to a grid of rectangular blocks. Consider a reservoir with an arbitrary areal configuration as shown in Fig. 6.1. Now superimpose on this map the smallest possible rectangle that totally encompasses the reservoir. A grid consisting of orthogonal lines in each of the coordinate directions is then constructed forming a collection of rectangles bounded by the larger rectangle. Each smaller rectangle is called a *cell* or *block*. The blocks interior to the wiggly line in Fig. 6.2 are referred to as *active* blocks while those outside are *inactive*. Calculations are performed only for the active blocks in a reservoir simulator.

It is not necessary that the grid lines be equally spaced, i.e., block sizes need not be uniform. There are two basic grid types employed in reservoir work, block-centered grids and lattice-centered grids. For simplicity, we discuss these in a 2-D environment, but the same ideas apply to 3-D systems. In a block-centered grid, a point, P, is associated with the grid block center having indices (i,j), while the interfaces carry $(i \pm \frac{1}{2}$,

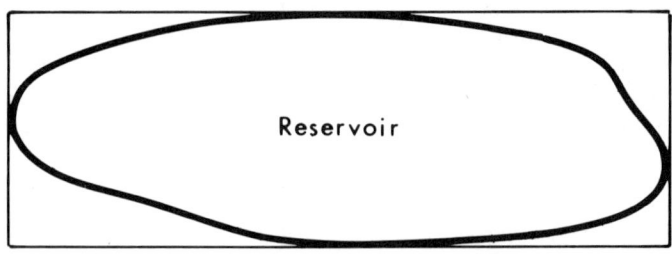

Fig. 6.1 Plan View of a Reservoir.

Single-Phase Multidimensional Flow 81

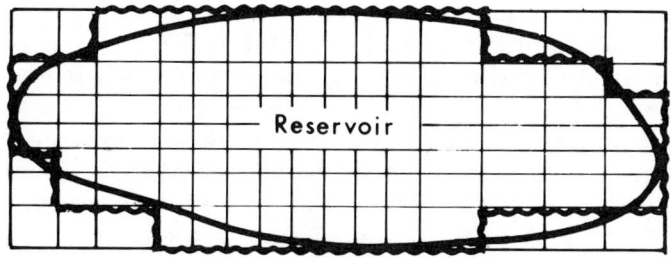

Fig. 6.2 Gridded Reservoir Map.

$j \pm \frac{1}{2}$) indices for a collection of blocks $i = 1, 2, \ldots, N_x$ in the x-direction, and $j = 1, 2, \ldots, N_y$ in the y-direction. In lattice-centered spacing, a point (i,j) is at the intersection of each grid line. Again, the grid is specified by the sequences, $i = 0, 1, 2, \ldots, N_x$ in the x-direction and $j = 0, 1, 2, \ldots, N_y$ in the y-direction. Block centers are at the half values for both indices. In Fig. 6.3 we illustrate block- and lattice-centered grids in 1-D.

The finite different representations of a system of partial differential equations is independent of the grid system employed; i.e., they are identical for both block- and lattice-centered grids. The only difference arises in the treatment of boundary conditions. The reason is the lattice-centered grid has boundaries coincident with the exterior calculation points, while the block-centered system has its boundaries one-half grid block beyond the exterior calculation points. Most commercial reservoir simulators employ block-centered grids, although in some applications, a lattice-type grid is preferable. For our purposes, it is sufficient to confine ourselves to block-centered grids. In this context we consider Dirichlet and Neumann-type boundary conditions. A Dirichlet boundary condition requires that the dependent variable be specified on the boundary. For example, a constant pressure (or temperature, etc.) boundary is a Dirichlet condition. On the other hand, if we specify the gradient of the dependent variable on the boundary, then we have a Neumann condition there. The most frequently

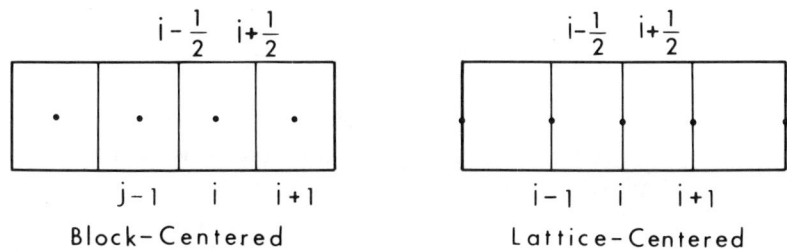

Fig. 6.3 Block- and Lattice-Centered Grid Systems.

employed Neumann condition in reservoir simulation is the "no-flow boundary;" i.e., if Φ is the flow potential, then $\nabla \Phi = 0$ on ∂R where ∂R is the closed boundary denoted by the wiggly line in Fig. 6.2. It is not unusual to have *mixed boundary conditions*, a Dirichlet condition over one part of ∂R and a Neumann condition over another. Reservoirs subject to water influx from an adjoining aquifer are examples. The water influx rate may be sufficient to maintain the potential at an essentially constant value in that part of the reservoir interfacing with the aquifer. This leads to a Dirichlet boundary condition there. Other parts of the reservoir may be sealed because of structural or stratigraphic trapping. Consequently, these represent no-flow boundaries of the Neumann type.

For block-centered grids, a Dirichlet condition is accounted for by simply specifying the value of the dependent variable (frequently pressure) at the block centers along that portion of the boundary affected. In other words, we neglect the fact that the block centers are not actually coincident with the reservoir boundaries. This approximation, of course, is improved if we use smaller grid spacings, i.e., a *refined grid*. However, this is not always practical in large reservoir problems. Furthermore, in global reservoir problems, efforts to do a better job on the boundaries are largely wasted since what goes on at interior points is little affected by boundary phenomena if the boundaries are sufficiently far removed. Indeed, in many instances, it makes no difference whether we specify a Neumann or Dirichlet condition. Where it does, we can always arbitrarily move the boundaries out to where their effects are not felt in the simulator, and concentrate on the areas of interest in the reservoir. The remarks above do not apply when we are examining single-well phenomena such as well-coning, etc., in a localized area. In such cases, realistic handling of boundary conditions can be very important and may entail a departure from block-centered schemes. Such considerations are treated by Settari and Aziz.[1]

A Neumann boundary condition for a block-centered grid system can be handled in one of two ways by: (1) using image blocks or (2) setting certain boundary coefficients to zero. Suppose for some 2-D region consisting of $N_x \times N_y$ blocks where $1 \leq i \leq N_x$ and $1 \leq j \leq N_y$, we require that $\nabla \Phi = 0$ on the far boundary in the x-direction, i.e., $\partial \Phi / \partial x = 0 \; \forall \; y$ there. Using a central difference to express this leads to

$$\Phi_{N_x-1,j} = \Phi_{N_x+1,j}, \; 1 \leq j \leq N_y$$

where additional blocks with centers at $(N_x + 1, j)$ are introduced. Such blocks are *image blocks* occurring outside the reservoir, but are so-called because they reflect the value of the dependent variable at the interior points $(N_x - 1, j)$. More frequently, in reservoir simulation, we treat expressions of the form $\nabla \cdot \{K \nabla \Phi\}$ where K is a spatially- and possibly time-dependent coefficient. If we require $\nabla \Phi = 0$ on ∂R we can, in this case, accomplish the same effect by setting $K = 0$ on ∂R. The nice thing about this approach is that one can specify flow or no-flow conditions by simply manipulating input data.

6.3 Slightly Compressible Flow

For a slightly compressible fluid in a homogeneous, isotropic, 2-D cartesian region where gravity is neglected, we have (see section 4.7)

$$\frac{\partial^2 p}{\partial X^2} + \frac{\partial^2 p}{\partial Y^2} = \frac{1}{\alpha}\frac{\partial p}{\partial t}; \quad \alpha \equiv \frac{k}{\phi\mu c} \tag{6.1}$$

which when normalized yields

$$u_{xx} + u_{yy} = u_t, \; t > 0, \; (x,y) \in R \tag{6.2}$$

where $R = [0, 1] \times [0, 1]$. On the far boundaries of R, i.e., at $x = 1 \; 0 \leq y \leq 1$ and $y = 1$, $0 \leq x \leq 1$ we invoke a Dirichlet condition, $u = 0$, while on the near boundaries $x = 0$, $0 \leq y \leq 1$ and $y = 0$, $0 \leq x \leq 1$, Neumann conditions are specified, $u_x = 0$ and $u_y = 0$, respectively. We also require an initial condition, $u = u_0$ say, at $t = 0$ to get a well-posed problem. Using the difference operators defined in chapter 5 and subscripting the x- and y-parts, the explicit formulation of Eq. 6.2 is

$$(\Delta_x^2 u^n)/\Delta x^2 + (\Delta_y^2 u^n)/\Delta y^2 = (\Delta_t u)/\Delta t. \tag{6.3}$$

A stability analysis indicates that $\Delta t(\Delta x^{-2} + \Delta y^{-2})$ should be $\leq \frac{1}{2}$ which is too restrictive for practical reservoir problems. If we appeal to the Crank-Nicolson scheme, then

$$\frac{\gamma_1}{2}\{\Delta_x^2 u^{n+1} + \Delta_x^2 u^n\} + \frac{\gamma_2}{2}\{\Delta_y^2 u^{n+1} + \Delta_y^2 u^n\} = \Delta_t u \tag{6.4}$$

where $\gamma_1 = \Delta t/\Delta x^2$ and $\gamma_2 = \Delta t/\Delta y^2$. This leads to a matrix problem of the type in Eq. 5.23; however, the coefficient matrix is *pentadiagonal* rather than tridiagonal. We defer treatment of such matrix problems until later. The matrix will be of order $N_x N_y$ which, for small computers, may be impractical to handle. An alternative approach is the fractional step method of Peaceman and Rachford[2] known as the Alternating Direction Implicit (ADI) procedure. The central idea is to consider a multidimensional problem as a collection of one-dimensional problems, each of which is solved over a fraction of a time step.† The associated matrix problems are always tridiagonal. This is discussed next.

6.4 Alternating Direction Implicit Procedure

Instead of Eq. 6.3, we express one of the coordinate directions implicitly leaving the other explicit and consider that time is advanced over half a time step. Then the roles of the implicit and explicit parts are interchanged to complete the time step. We refer to the spatial derivative evaluated implicitly as the *sweeping direction*. Thus, if the x-part is implicit,

$$(\Delta_x^2 u^{n+1/2})/\Delta x^2 + (\Delta_y^2 u^n)/\Delta y^2 = \frac{2}{\Delta t}(u^{n+1/2} - u^n) \tag{6.5}$$

for the x-sweep, and

$$(\Delta_x^2 u^{n+1/2})/\Delta x^2 + (\Delta_y^2 u^{n+1})/\Delta y^2 = \frac{2}{\Delta t}(u^{n+1} - u^{n+1/2}) \tag{6.6}$$

for the y-sweep. Eqs. 6.5 and 6.6 lead to matrix problems of the form

$$\mathbf{H}\mathbf{u}^{n+1/2} = \mathbf{d}_x \tag{6.7}$$

$$\mathbf{V}\mathbf{u}^{n+1} = \mathbf{d}_y \tag{6.8}$$

where both $\mathbf{H}$ and $\mathbf{V}$ are tridiagonal matrices. In $\mathbf{H}$ the diagonal term is $2(1 + \gamma_1)$ while the off-diagonals are $-\gamma_1$. In $\mathbf{V}$ the diagonal and off-diagonals are $2(1 + \gamma_2)$ and $-\gamma_2$, respectively. Also $\mathbf{d}_x$ contains all terms in Eq. 6.5 evaluated at time level n while $\mathbf{d}_y$ contains those in Eq. 6.6 at time-level $n + \frac{1}{2}$. The solution process is started using the initial condition, $u = u_0$ at time level zero, i.e., $n = 0$, to determine the right-hand side of Eq. 6.7. Solution of Eq. 6.7 for $\mathbf{u}^{n+1/2}$ permits us to evaluate $\mathbf{d}_y$ for use in the y-sweep. Consequently, alternately solving Eqs. 6.7 and 6.8 repeatedly advances the solution in time.

A more realistic application of ADI is to single-phase flow in anisotropic, heterogeneous media containing source (or sink) terms. Consider the problem

$$\frac{\partial}{\partial x}\left(\frac{k_x}{\mu}\frac{\partial p}{\partial x}\right) + \frac{\partial}{\partial y}\left(\frac{k_y}{\mu}\frac{\partial p}{\partial y}\right) + f(x,y,t) = \phi c \frac{\partial p}{\partial t} \tag{6.9}$$

† This concept has been exploited in contexts other than finite difference methods.[3-6]

Single-Phase Multidimensional Flow

where $(x,y) \in R$, $t > 0$. On R, we define partitions

$$\Pi_x: 0 < x_1 < x_2 < x_3 < \ldots < x_{N_x}$$

$$\Pi_y: 0 < y_1 < y_2 < y_3 < \ldots < y_{N_y}$$

such that a collection of rectangular blocks is defined, not necessarily equally spaced. If the grid is block-centered, then

$$x_i = \tfrac{1}{2}(x_{i-1/2} + x_{i+1/2}),\ \Delta x_i = x_{i+1/2} - x_{i-1/2},\ i = 1, 2, \ldots, N_x$$
$$y_j = \tfrac{1}{2}(y_{j-1/2} + y_{j+1/2}),\ \Delta y_j = y_{j+1/2} - y_{j-1/2},\ j = 1, 2, \ldots, N_y$$

A second order correct approximation for the derivatives in Eq. 6.9 is

$$\frac{\partial}{\partial s}\left(a_s \frac{\partial p}{\partial s}\right) = \frac{(2a_s)_{k+1/2}\,(p_{k+1} - p_k)}{\Delta s_k(\Delta s_{k+1} + \Delta s_k)} - \frac{(2a_s)_{k-1/2}\,(p_k - p_{k-1})}{\Delta s_k(\Delta s_k + \Delta s_{k-1})}$$

where $s = x$ or y, $k = i$ or j and $a_s = k_x/\mu$ or k_y/μ.

Define $(A_s)_{k \pm 1/2} \equiv (2a_s)_{k \pm 1/2}/(\Delta s_k + \Delta s_{k \pm 1})$ then,

$$\frac{\partial}{\partial s}\left(a_s \frac{\partial p}{\partial s}\right) = \{(A_s)_{k+1/2}(p_{k+1} - p_k) - (A_s)_{k-1/2}(p_k - p_{k-1})\}/\Delta s_k$$
$$\equiv \Delta_s A_s \Delta_s p / \Delta s_k. \tag{6.10}$$

With the difference notation in Eq. 6.10, the finite difference equation for Eq. 6.9 is

$$(\Delta_x A_x \Delta_x p)/\Delta x_i + (\Delta_y A_y \Delta_y p)/\Delta y_j + f(x,y,t) = \frac{\phi c}{\Delta t}(p_{ij}^{n+1} - p_{ij}^n). \tag{6.11}$$

If we multiply Eq. 6.11 by $\Delta x_i \Delta y_j$; then,

$$\Delta_x T_x \Delta_x p + \Delta_y T_y \Delta_y p + q_{ij} = \frac{\alpha}{\Delta t}(p_{ij}^{n+1} - p_{ij}^n) \tag{6.12}$$

where $\alpha \equiv c \Delta x_i \Delta y_j \phi$; $q_{ij} \equiv \Delta x_i \Delta y_j f(x,y,t)$,

$$\Delta_x T_x \Delta_x p \equiv (T_x)_{i+1/2}\,(p_{i+1,j} - p_{i,j}) - (T_x)_{i-1/2}\,(p_{i,j} - p_{i-1,j}),$$

$(T_x)_{i+1/2} \equiv (A_x)_{i+1/2} \Delta y_j$, and $(T_y)_{j-1/2} \equiv (A_y)_{j-1/2} \Delta x_i$, i.e.,

$$(T_x)_{i+1/2} \equiv \frac{2(k_x/\mu)_{i+1/2} \Delta y_j}{\Delta x_i + \Delta x_{i+1}},\ \text{for example.}$$

The T's in the above expressions are called *transmissibilities* or *transmissivities*. (The latter term is most frequently used by hydrologists, while the former is commonly used in the oil industry, perhaps incorrectly.[7]) One should observe the analogous notation in Eqs. 6.9 and 6.12; i.e., the difference operator for $\partial/\partial x$ becomes Δ_x and similarly $\partial/\partial y$ has the analog Δ_y. Thus, if $k_x/\mu = k_y/\mu = $ a constant, we have

$$\Delta_x^2 p + \Delta_y^2 p + q_{ij}^* = \frac{\beta}{\Delta t}(p_{ij}^{n+1} - p_{ij}^n) \qquad (6.13)$$

for the linear problem where $\beta \equiv (\phi\mu c/k)(\Delta x_i \Delta y_j)$, and $q_{ij}^* \equiv q_{ij}\mu/k$. Frequently we use the operator $\Delta T \Delta p$ to denote the spatial difference operators in 1-, 2-, or 3-D. Thus it can represent the sum of the two left-hand side terms in Eq. 6.12. In 3-D it includes an additional term similar to those above for the z-direction. Thus a compact form for Eq. 6.12 is

$$\Delta T \Delta p + q_{ij} = \frac{\alpha}{\Delta t} \Delta_t p$$

where $\Delta_t p$ denotes the time difference in Eq. 6.12.

The ADI formulation of Eq. 6.9 is

$$x\text{-sweep: } \Delta_x T_x \Delta_x p^{n+1/2} + \Delta_y T_y \Delta_y p^n + q^n = \frac{2\alpha}{\Delta t}(p^{n+1/2} - p^n) \qquad (6.14)$$

$$y\text{-sweep: } \Delta_x T_x \Delta_x p^{n+1/2} + \Delta_y T_y \Delta_y p^{n+1} + q^n = \frac{2\alpha}{\Delta t}(p^{n+1} - p^{n+1/2}) \qquad (6.15)$$

where indices i and j on the q-terms and the right-hand sides are suppressed. Eq. 6.14 can be written as

$$-a_i p_{i-1}^{n+1/2} + b_i p_i^{n+1/2} - c_i p_{i+1}^{n+1/2} = d_i \qquad (6.16)$$

where $a_i = (T_x)_{i-1/2}$, $b_i = \left[(T_x)_{i+1/2} + (T_x)_{i-1/2} + \frac{2\alpha}{\Delta t}\right]$, $c_i = (T_x)_{i+1/2}$,

and $d_i = (T_y)_{j-1/2} p_{j-1}^n - \left[(T_y)_{j+1/2} + (T_y)_{j+1/2} - \frac{2\alpha}{\Delta t}\right] p_{ij}^n + (T_y)_{j+1/2} p_{j+1} + q_{ij}^n.$

Eq. 6.15 becomes

$$-\hat{a}_j p_{j-1}^{n+1} + \hat{b}_j p_{ij}^{n+1} - \hat{c}_j p_{j+1}^{n+1} = \hat{d}_j \qquad (6.17)$$

where $\hat{a}_j = (T_y)_{j-1/2}$, $\hat{b}_j = \left[(T_y)_{j-1/2} + (T_y)_{j+1/2} + \dfrac{2\alpha}{\Delta t} \right]$, $\hat{c}_j = (T_y)_{j+1/2}$,

and $\hat{d}_j = (T_x)_{i-1/2} p_{i-1}^{n+1/2} - \left[(T_x)_{i-1/2} + (T_x)_{i+1/2} - \dfrac{2\alpha}{\Delta t} \right] p_{ij}^{n+1/2}$
$\qquad\qquad\qquad\qquad\qquad\qquad + (T_x)_{i+1/2} p_{i+1}^{n+1/2} + q_{ij}^{n+1/2}.$

In the expressions above, with the exception of terms referring to the point (i,j), only the "fastest moving" index is shown to avoid a clutter of subscripts. By fastest moving index, we mean the index that runs through all its values for *fixed* values of the others. For example, for the *x*-sweep, *i* is the fastest moving index since for a fixed j, $i = 1, 2, \ldots, N_x$. For simplicity of notation we follow this practice throughout. Note that both Eqs. 6.16 and 6.17 lead to tridiagonal matrix problems for each sweep as anticipated. Unlike the matrix problems associated with Eqs. 6.5 and 6.6, however, the matrix coefficients in this case are not constants but are space-dependent functions. Later we shall see that the treatment of multiphase flow problems involves space and time-dependent matrix coefficients.

The ADI procedure of Peaceman and Rachford is unconditionally stable as will be shown. However, truncation errors can become prohibitive for large Δt which severely restricts the time-step size in many practical applications. Furthermore, when the transmissibilities are strongly contrasting in each of the coordinate directions, convergence is difficult if not impossible to achieve. Again, in treating some problems involving moving fronts with ADI, a substantial "ringing effect" is observed in the vicinity of the front. This produces oscillations in the dependent variable about some fixed value. These oscillations are not instabilities in the true sense of the word, since they usually dampen with time. Actually, they are associated with "Crank-Nicolson noise" insofar as Peaceman-Rachford ADI is a perturbed form of the Crank-Nicolson procedure (see section 6.6). There are techniques for controlling the oscillations[8] and even removing them entirely by smoothing procedures.[9] Nevertheless, the other disadvantages of ADI are such that it is now rarely, if ever, used in commercial simulators. Some, however, provide an option to use an iterative ADI approach (IADI) described next.

6.5 Iterative Alternating Direction Implicit Method

The procedure we describe is that of Peaceman and Rachford.[2] Later we discuss variations of this technique. Instead of treating each sweep as an advancement in time level, one can express the pertinent equations as advancements in *iteration levels* to get from n to $n + 1$. Thus, Eqs. 6.14 and 6.15 can be written as

$$\Delta_x T_x \Delta_x p^{k+1/2} + \Delta_y T_y \Delta_y p^k - \frac{2\alpha}{\Delta t} p^{k+1/2} = -\frac{2\alpha}{\Delta t} p^n - q^n \quad (6.18)$$

$$\Delta_x T_x \Delta_x p^{k+1/2} + \Delta_y T_y \Delta_y p^{k+1} - \frac{2\alpha}{\Delta t} p^{k+1} = -\frac{2\alpha}{\Delta t} p^n - q^n \quad (6.19)$$

Note in Eq. $6.19 - 2\alpha/\Delta t\ p^n$ appears rather than $-2\alpha/\Delta t\ p^{n+1/2}$ as in Eq. 6.15. To compensate for this slight modification, an appropriate multiple of the difference in iterates is added to each side of Eqs. 6.18 and 6.19 to arrive at the iterative finite difference scheme:

$$\Delta_x T_x \Delta_x p^{k+1/2} + \Delta_y T_y \Delta_y p^k - \frac{2\alpha}{\Delta t} p^{k+1/2} = H_k(p^{k+1/2} - p^k) - C \quad (6.20)$$

$$\Delta_x T_x \Delta_x p^{k+1/2} + \Delta_y T_y \Delta_y p^{k+1} - \frac{2\alpha}{\Delta t} p^{k+1} = H_k(p^{k+1} - p^{k+1/2}) - C \quad (6.21)$$

where H_k is a product of an iteration parameter† and a normalizing factor and $C = 2\alpha/\Delta t\ p^n + q^n$. If in Eqs. 6.20 and 6.21, the implicit terms are placed on the left-hand sides, and the explicit terms on the right-hand sides, then again the problem takes on a tridiagonal form.

The normalizing factor used to formulate H_k is the sum of the transmissibilities around the four faces of an (i,j)-block. Thus we write $H_k = \sigma_k (\Sigma T)$ where $\Sigma T \equiv (T_x)_{i+1/2,j} + (T_x)_{i-1/2,j} + (T_y)_{i,j+1/2} + (T_y)_{i,j-1/2}$ and σ_k is an iteration parameter. The subscript, k, is assigned to σ since normally it will change from iteration to iteration. In practice, a set of parameters are employed, $\sigma_1, \sigma_2, \ldots, \sigma_K$, where $\sigma_1 < \sigma_2 < \ldots < \sigma_K$ and one cycles through them from the smallest to the largest (or sometimes vice versa) and repeats the cycle, in whole or in part, until convergence is achieved. The parameters should be geometrically spaced, i.e., $\sigma_{k+1}/\sigma_k = \alpha$, a constant. If a total of K parameters are chosen per cycle, then $\sigma_K/\sigma_1 = \alpha^{K-1}$. If σ_1, σ_K (the minimum and maximum parameters) and the number of parameters per cycle are known, then α may be calculated from

$$\ln(\alpha) = \frac{\ln(\sigma_K/\sigma_1)}{K-1}. \quad (6.22)$$

---

† Iteration parameters are real numbers used to accelerate convergence rates.

Usually K is 4 or 5 for a small range on the σ_k (e.g., .01 to 2) and 6 to 8 for a large range (e.g., .0001 to 2).

Unfortunately, there is no theoretical basis for selecting iteration parameters for many practical simulation problems, except for highly idealized cases. For example, if N_x and N_y are sufficiently large and the transmissibilities T_x and T_y are uniform, but not necessarily equal, then the minimum parameter is given by

$$\sigma_1 = \text{Min} \left\{ \frac{\pi^2}{2N_x^2}\left(\frac{1}{1+T_y/T_x}\right), \frac{\pi^2}{2N_y^2}\left(\frac{1}{1+T_x/T_y}\right) \right\}. \qquad (6.23)$$

If $T_x = T_y$, the maximum is 1 and if $T_x \gg T_y$ or $T_y \gg T_x$, the maximum is 2. For equally spaced grid points, $T_x = k_x \Delta y/\Delta x$, $T_y = k_y \Delta x/\Delta y$, then

$$\sigma_1 = \min \left\{ \frac{\pi^2}{2N_x^2} \frac{1}{1+\frac{k_y \Delta x^2}{k_x \Delta y^2}}, \frac{\pi^2}{2N_y^2} \frac{1}{1+\frac{k_x \Delta y^2}{k_y \Delta x^2}} \right\}. \qquad (6.24)$$

For an areal (2-D) problem, generally $\Delta x \approx \Delta y$ and if $k_x = k_y$, then the two terms in Eq. 6.24 are equal. For cross-sectional problems, however, generally $N_x \gg N_y$, $\Delta x \gg \Delta y$ and the first term is smaller.

In cases where the transmissibilities are highly variable, Eqs. 6.23 and 6.24 can serve only as rough guides. One could conceivably compute σ_1 for each (i,j)-block using Eq. 6.23 and then select the minimum from that set. The maximum value will always be between 2 and a number slightly less than 1.[10] The convergence rate is usually insensitive to its actual magnitude within this range. However, there is extreme sensitivity to σ_1 for some problems. A change from 0.005 to 0.0001 may be critical and cause divergence. When this occurs 3 or 4 trial runs may be necessary to determine a σ_k set that will be satisfactory.

Frequently as convergence is approached, the difference between successive iterates becomes so small that significant digits are cancelled on a computer. This occurs sooner on machines with short word length. This error can be minimized by solving for the displacement in each iteration rather than for the iterate itself. Thus if we define $PX \equiv p^{k+1/2} - p^k$ and $PY \equiv p^{k+1} - p^k$ then, Eqs. 6.20 and 6.21 become

$$\Delta_x T_x \Delta_x PX - \left(\frac{2\alpha}{\Delta t} + H_k\right) PX = \frac{2\alpha}{\Delta t} p^k - \Delta T \Delta p^k - C \qquad (6.25)$$

and, $\Delta_y T_y \Delta_y PY - \left(\frac{2\alpha}{\Delta t} + H_k\right) PY = \frac{2\alpha}{\Delta t} p^k - \Delta T \Delta p^k$
$$- C - \Delta_x T_x \Delta_x PX - H_k PX. \qquad (6.26)$$

Let $G \equiv 2\alpha/\Delta t\, p^k - \Delta T\Delta p^k - C$ and notice that the right-hand side of Eq. 6.26 can be expressed in terms of PX and G. Since $G = \Delta_x T_x \Delta_x PX - (2\alpha/\Delta t + H_k)PX$ from Eq. 6.25, we get the residual form,

$$x\text{-sweep:}\quad \Delta_x T_x \Delta_x PX - \left(\frac{2\alpha}{\Delta t} + H_k\right) PX = G \tag{6.27}$$

$$y\text{-sweep:}\quad \Delta_y T_y \Delta_y PY - \left(\frac{2\alpha}{\Delta t} + H_k\right) PY = -2\left(\frac{\alpha}{\Delta t} + H_k\right) PX \tag{6.28}$$

Now consider the implicit finite difference formulation of the original problem:

$$\Delta T\Delta p^{n+1} + q^n = \frac{2\alpha}{\Delta t}(p^{n+1} - p^n). \tag{6.29}$$

Since $C = q^n + 2\alpha/\Delta t\, p^n$, then we have

$$-\frac{2\alpha}{\Delta t} p^{n+1} + \Delta T\Delta p^{n+1} + C = 0. \tag{6.30}$$

Notice G is identical to Eq. 6.30 using the kth iterate. If convergence to the exact solution at time level $n+1$ is achieved then G would be equal to zero. This observation can be used as a closure criterion. Let the residual, $R_{i,j}$ be given by

$$R_{ij} = \Delta T\Delta p^k - \frac{2\alpha}{\Delta t} p^k + C, \tag{6.31}$$

and invoke the following closure criteria:

(1) $\epsilon_1 = \sum\limits_{i,j} |R_{ij}|$
$\qquad = \sum\limits_{i=1}^{N_x} \sum\limits_{j=1}^{N_y} |R_{ij}|$ (absolute residual)

(2) $\epsilon_2 = \dfrac{\sum\limits_{i,j} |R_{ij}|}{\sum\limits_{i,j} |Q_{ij}|}$ (normalized residual)

Usually $\epsilon_1 \leq 0.001$ and $\epsilon_2 \leq .05$ are satisfactory for closure.

6.6 Stability and Accuracy of ADI Methods

We concern ourselves here only with the stability and truncation error of the noniterative ADI methods of Peaceman-Rachford. The iterative procedure will generally display the same stability and accuracy characteristics as the direct method.

Because of the spatial variation of permeability in Eq. 6.9, a stability analysis is difficult to achieve without some simplification. For this reason, attention is directed to a von Neumann stability analysis of Eq. 6.2 to illustrate the ideas involved. Let the error component be given by

$$\epsilon_{ij}^n = \zeta^n e^{Jp i \Delta x} e^{Jq j \Delta y} \tag{6.32}$$

analogous to Eq. 5.34 where $p = k\pi/N_x\Delta x$ and $q = k\pi/N_y\Delta y$, $k = 1, 2, 3, \ldots$, and we require that $|\zeta| \leq 1$ for stability. Since the error component in Eq. 6.32 is propagated by the difference equations (Eqs. 6.5 and 6.6), substitution of ϵ_{ij} into the latter two equations leads to

$$\zeta = \frac{1 - 4\gamma \sin^2(q\Delta y/2)}{1 + 4\gamma \sin^2(p\Delta x/2)} \tag{6.33}$$

for the x-sweep, and

$$\zeta = \frac{1 - 4\gamma \sin^2(p\Delta x/2)}{1 + 4\gamma \sin^2(q\Delta y/2)} \tag{6.34}$$

for the y-sweep. In arriving at Eqs. 6.33 and 6.34 each sweep is considered as a full step in time; i.e., we omitted the 2's in Eqs. 6.5 and 6.6, and assumed the x-sweep takes us from level n to $n + 1$, and the y-sweep from $n + 1$ to $n + 2$. Also, $\gamma = \Delta t/\Delta x^2 = \Delta t/\Delta y^2$. When $\gamma > \frac{1}{2}$, $p = 1$, $q = N_y - 1$ then $|\zeta| > 1$, i.e., Eq. 6.5, becomes unstable. Similarly, Eq. 6.6 is unstable for certain values of the same parameters. We demonstrate that $|\zeta| > 1$ for Eq. 6.33 only.

Consider some function $f(x,\hat{x}) = (1 - x)/(1 + \hat{x})$ where $\hat{x}$ is fixed. Then if $0 < x_1 \leq x \leq x_2$, it follows that $f_x < 0 \; \forall \; x \in [x_1, x_2]$ and this implies that f_{max} occurs at one of the end-points, i.e.,

$$f_{max} = \max\left\{\left|\frac{1 - x_1}{1 + \hat{x}}\right|, \left|\frac{1 - x_2}{1 + \hat{x}}\right|\right\}.$$

In Eq. 6.33, set $\gamma = 1$ and suppose $\Delta x = \Delta y = \pi/N$ where $N_x = N_y \equiv N$; thus,

$$\zeta = \frac{1 - 4\sin^2\left(\frac{k\pi}{2N}\right)}{1 + 4\sin^2\left(\frac{\pi}{2N}\right)}, \quad k = 1, 2, \ldots, N-1 \qquad (6.35)$$

where $N > 1$. Now $x_1 = \hat{x} = 4\sin^2(\pi/2N)$ and

$$x_2 = 4\sin^2\left[\frac{\pi(N-1)}{2N}\right] = 4\cos^2\left(\frac{\pi}{2N}\right).$$

Consequently,

$$|\zeta|_{\max} = \max\left(\left|\frac{1 - 4\sin^2(\pi/2N)}{1 + 4\sin^2(\pi/2N)}\right|, \left|\frac{1 - 4\cos^2(\pi/2N)}{1 + 4\sin^2(\pi/2N)}\right|\right). \qquad (6.36)$$

Moreover, $\cos^2(\pi/2N) \geq \sin^2(\pi/2N) \; \forall \; N > 1$; thus, $|\zeta|_{\max}$ is the second term in the curly brackets in Eq. 6.36 and this is greater than one for N sufficiently large.

If we consider Eqs. 6.33 and 6.34 as the results of half-steps in time, then the right-hand side of Eq. 6.33 $= \zeta^{n+1/2}/\zeta^n$ and the right-hand side of Eq. 6.34 $= \zeta^{n+1}/\zeta^{n+1/2}$ such that

$$\zeta = \frac{1 - 2\gamma \sin^2(q\Delta y/2)}{1 + 2\gamma \sin^2(p\Delta x/2)} \cdot \frac{1 - 2\gamma \sin^2(p\Delta x/2)}{1 + 2\gamma \sin^2(q\Delta y/2)}. \qquad (6.37)$$

One can readily see in this case that $|\zeta| \leq 1$ for $\gamma > 0$. Clearly, the individual sweeps of ADI can become unstable, while the composite effect of the x and y-sweeps yields an unconditionally stable procedure.

As an alternative to a truncation error analysis, we relate the ADI procedure to the Crank-Nicolson scheme via an *overall equation* as done by Douglas[11], and thereby determine the truncation error much more simply. For example, subtract Eq. 6.14 from Eq. 6.15 and get

$$\Delta_y T_y \Delta_y (p^{n+1} - p^n) = \frac{2\alpha}{\Delta t}(p^{n+1} + p^n - 2\, p^{n+1/2})$$

from which

$$p^{n+1/2} = \tfrac{1}{2}(p^{n+1} + p^n) - \frac{\Delta t}{4\alpha}\Delta_y T_y \Delta_y (p^{n+1} - p^n). \qquad (6.38)$$

Adding Eqs. 6.14 and 6.15 yields

$$2\Delta_x T_x \Delta_x p^{n+1/2} + \Delta_y T_y \Delta_y (p^{n+1} + p^n) + 2\, q^n = \frac{2\alpha}{\Delta t}(p^{n+1} - p^n). \qquad (6.39)$$

Single-Phase Multidimensional Flow   93

Now, substitute Eq. 6.38 in Eq. 6.39 and get the overall equation,

$$\Delta_x T_x \Delta_x \frac{(p^{n+1} + p^n)}{2} + \Delta_y T_y \Delta_y \frac{(p^{n+1} + p^n)}{2} + q^n$$

$$= \frac{\alpha}{\Delta t}(p^{n+1} - p^n) + \frac{\Delta t}{4\alpha} \Delta_x T_x \Delta_x \Delta_y T_y \Delta_y (p^{n+1} - p^n) \quad (6.40)$$

which is equivalent to the Crank-Nicolson approximation augmented by a perturbation term (the last term on the right-hand side). Consequently the error is $0(\Delta x^2 + \Delta y^2 + \Delta t^2)$ and convergence is guaranteed when $\Delta s \to 0$, $s = x, y$ and t.

6.7 Flow Problems in 3-D

An attempt to extend the Peaceman-Rachford technique to single-phase flow in 3-D, shows that the composite of three sweeps can lead to instabilities. For example, for the problem

$$u_{xx} + u_{yy} + u_{zz} = u_t \quad (6.41)$$

the *amplification factor*, ζ, as determined by a von Neumann analysis is

$$\zeta = \frac{(a+b+c)(b-3)}{(1+a)(1+b)(1+c)} + 1 \quad (6.42)$$

where

$$a = 4\gamma_1 \sin^2(p\Delta x/2); \gamma_1 = \Delta t/\Delta x^2$$

$$b = 4\gamma_2 \sin^2(q\Delta y/2); \gamma_2 = \Delta t/\Delta y^2$$

$$c = 4\gamma_3 \sin^2(r\Delta z/2); \gamma_3 = \Delta t/\Delta z^2$$

When $b > 3$, the system becomes unstable.

Douglas and Rachford[12] developed an unconditionally stable ADI scheme for 3-D which we illustrate for Eq. 6.41. The formulation is

$$x: \quad \Delta_x^2 u^{n+1/3} + \Delta_y^2 u^n + \Delta_z^2 u^n = \frac{u^{n+1/3} - u^n}{\Delta t} \quad (6.43)$$

$$y: \quad \Delta_y^2 u^{n+2/3} = \Delta_y^2 u^n + (u^{n+2/3} - u^{n+1/3})/\Delta t \quad (6.44)$$

$$z: \quad \Delta_z^2 u^{n+1} = \Delta_z^2 u^n + (u^{n+1} - u^{n+2/3})/\Delta t \quad (6.45)$$

for each of the three sweeps. If we solve Eq. 6.44 for $\Delta_y^2 u^n$ and substitute into Eq. 6.43, we get an alternative expression for the y-sweep. Similarly, if we use Eq. 6.45 to replace $\Delta_z^2 u^n$ in the new y-sweep equation, we get a new expression for the z-sweep. Thus, the alternate forms are

$$x: \quad \Delta_x^2 u^{n+1/3} + \Delta_y^2 u^n + \Delta_z^2 u^n = (u^{n+1/3} - u^n)/\Delta t \qquad (6.46)$$

$$y: \quad \Delta_x^2 u^{n+1/3} + \Delta_y^2 u^{n+2/3} + \Delta_z^2 u^n = (u^{n+2/3} - u^n)/\Delta t \qquad (6.47)$$

$$z: \quad \Delta_x^2 u^{n+1/3} + \Delta_y^2 u^{n+2/3} + \Delta_z^2 u^{n+1} = (u^{n+1} - u^n)/\Delta t \qquad (6.48)$$

Eqs. 6.46–6.48 differ from the Peaceman-Rachford formulation only in the right-hand sides. In the equations above, the time displacements are measured from u^n in all three sweeps rather than from $u^{n+1/3}$ and $u^{n+2/3}$ in the y- and z-sweeps, respectively, of the Peaceman-Rachford method. Also, division is by the total time step, Δt.

This technique can be applied to 2-D problems, however, the time truncation error is greater. The overall equation is, in this case

$$(\Delta_x^2 + \Delta_y^2) u^{n+1} = (u^{n+1} - u^n)/\Delta t + \frac{\Delta t}{2} \Delta_x^2 \Delta_y^2 (u^{n+1} - u^n). \qquad (6.49)$$

Consequently, the truncation error is $0(\Delta x^2 + \Delta y^2 + \Delta t)$. Clearly, the Peaceman-Rachford method is superior for 2-D problems.

An iterative Douglas-Rachford scheme for the 3-D problem analogous to Eq. 6.9 has the form

$$x: \Delta_x T_x \Delta_x p^{k+1/3} + \Delta_y T_y \Delta_y p^k + \Delta_z T_z \Delta_z p^k - \frac{\alpha}{\Delta t} p^{k+1/3}$$
$$= H_k(p^{k+1/3} - p^k) - C \qquad (6.50)$$

$$y: \Delta_x T_x \Delta_x p^{k+1/3} + \Delta_y T_y \Delta_y p^{k+2/3} + \Delta_z T_z \Delta_z p^k - \frac{\alpha}{\Delta t} p^{k+2/3}$$
$$= H_k(p^{k+2/3} - p^k) - C \qquad (6.51)$$

$$z: \Delta_x T_x \Delta_x p^{k+1/3} + \Delta_y T_y \Delta_y p^{k+2/3} + \Delta_z T_z \Delta_z p^{k+1} - \frac{\alpha}{\Delta t} p^{k+1}$$
$$= H_k(p^{k+1} - p^k) - C \qquad (6.52)$$

where $\alpha \equiv \phi c \Delta x_i \Delta y_j \Delta z_k$ and $C \equiv (\alpha/\Delta t) p^n + q^n$. This, like the direct Douglas-Rachford method, is unconditionally stable with truncation error of $0(\Delta x^2 + \Delta y^2 + \Delta z^2 + \Delta t)$, i.e., it is second order correct in space and first order correct in time.

To establish a relationship between Peaceman-Rachford ADI and Douglas-Rachford ADI, consider half a time step in the Douglas-Rachford method and extrapolate to the full time step. For example, the 2-D Douglas-Rachford formulation for Eq. 6.2 is

$$\Delta_x^2 u^{n+1/2} + \Delta_y^2 u^n = \frac{u^{n+1/2} - u^n}{\Delta t} \tag{6.53}$$

$$\Delta_y^2 u^{n+1} = \Delta_y^2 u^n + \frac{u^{n+1} - u^{n+1/2}}{\Delta t} \tag{6.54}$$

If we make the replacements, $u^{n+1/4} \rightarrow u^{n+1/2}$, $u^{n+1/2} \rightarrow u^{n+1}$ and $\Delta t/2 \rightarrow \Delta t$ then,

$$\Delta_x^2 u^{n+1/4} + \Delta_y^2 u^n = \frac{(u^{n+1/4} - u^n)}{\Delta t/2} \tag{6.55}$$

$$\Delta_y^2 u^{n+1/2} = \Delta_y^2 u^n + \frac{u^{n+1/2} - u^{n+1/4}}{\Delta t/2} \tag{6.56}$$

The extrapolation to the full time step is done using

$$u^{n+1} = u^n + 2(u^{n+1/2} - u^n) \tag{6.57}$$

from which we get

$$u^{n+1/2} = (u^{n+1} + u^n)/2 \tag{6.58}$$

$$u^{n+1/2} - u^n = (u^{n+1} - u^n)/2 \tag{6.59}$$

Now substitute Eqs. 6.58 and 6.59 into Eq. 6.49 to get

$$(\Delta_x^2 + \Delta_y^2)(u^{n+1} + u^n)/2 = (u^{n+1} - u^n)/\Delta t + \frac{\Delta t}{4} \Delta_x^2 \Delta_y^2 (u^{n+1} - u^n) \tag{6.60}$$

which is the overall Peaceman-Rachford equation analogous to Eq. 6.40.

Brian[13] used this idea for a 3-D formulation that was both second order correct in space and time. Namely, he formulated the Douglas-Rachford method over half a time step and extrapolated to the $(n+1)^{st}$ time level. Consequently, for Eq. 6.41.

x:
$$\Delta_x^2 u^{n+1/6} + \Delta_y^2 u^n + \Delta_z^2 u^n = 2(u^{n+1/6} - u^n)/\Delta t \tag{6.61}$$

y:
$$\Delta_y^2 u^{n+1/3} = \Delta_y^2 u^n + 2(u^{n+1/3} - u^{n+1/6})/\Delta t \qquad (6.62)$$

z:
$$\Delta_y^2 u^{n+1/2} = \Delta_y^2 u^n + 2(u^{n+1/2} - u^{n+1/3})/\Delta t \qquad (6.63)$$

Again using Eqs. 6.57–6.59, we get

$$\Delta_x^2 + \Delta_y^2 + \Delta_z^2 (u^{n+1} + u^n)/2 = (u^{n+1} - u^n)/\Delta t + \text{perturbation term} \qquad (6.64)$$

which is a variation of Crank-Nicolson's formulation with truncation error of $0(\Delta x^2 + \Delta y^2 + \Delta z^2 + \Delta t^2)$. Douglas[14] suggested formulating the Crank-Nicolson procedure directly for the 3-D problem; i.e.,

x:
$$\Delta_x^2 (u^{n+1/3} + u^n)/2 + \Delta_y^2 u^n + \Delta_z^2 u^n = (u^{n+1/3} - u^n)/\Delta t \qquad (6.65)$$

y:
$$\Delta_x^2 (u^{n+1/3} + u^n)/2 + \Delta_y^2 (u^{n+2/3} + u^n)/2 + \Delta_z^2 u^n = (u^{n+2/3} - u^n)/\Delta t \qquad (6.66)$$

z: $\Delta_x^2 (u^{n+1/3} + u^n)/2 + \Delta_y^2 (u^{n+2/3} + u^n)/2 + \Delta_z^2 (u^{n+1} + u^n)/2$
$$= (u^{n+1} - u^n)/\Delta t \qquad (6.67)$$

Eqs. 6.65–6.67 lead to the same overall equation given in Eq. 6.64; thus, the truncation error is the same as Brian's method. The only difference between the two is that the computed intermediate values are different.

6.8 Band Matrix Problems

If the pressure in a reservoir remains essentially constant over a long period of time then the right-hand side of Eq. 6.9 is zero. From a practical point of view, such *steady-state* conditions exist only when the reservoir has been undisturbed (i.e., prior to drilling), at abandonment, under full pressure maintenance, or when the single-phase fluid is incompressible ($c = 0$). The resultant equation is then *elliptic* (see Appendix A.5.2). Any of the iterative ADI schemes discussed thus far can be applied to such a system. Consequently, a simulator that solves the 2 or 3-D parabolic problem given by Eqs. 6.20–6.21 and Eqs. 6.50–6.52 can equally well handle the companion elliptic problem that one obtains by setting $c = 0$. This is readily accomplished because c is usually specified as input data to the program. In so doing, the notion of taking a step in time vanishes, and we view the iterations as steps between different steady-state levels.

We now draw attention to direct methods of solution that depart from alternating direction schemes. This involves examining the structure of the matrices involved and developing an algorithm that capitalizes on their structures. Consider Eq. 6.12 in its expanded form

$$-(T_y)_{j-1/2}p_{j-1} - (T_x)_{i-1/2}p_{i-1} + \left(\Sigma T + \frac{\alpha}{\Delta t}\right)p_{ij} - (T_x)_{i+1/2}p_{i+1} - (T_y)_{j+1/2}p_{j+1}$$
$$= \frac{\alpha}{\Delta t}p_{ij}^n + q_{ij}^n \equiv \tilde{q}_{ij}. \quad (6.68)$$

where $\Sigma T = (T_x)_{i-1/2} + (T_x)_{i+1/2} + (T_y)_{j-1/2} + (T_y)_{j+1/2}$ as before.

In Eq. 6.68 we show only the fastest moving indices and suppress the superscript, $n+1$, on the left-hand side. To examine the matrix structure arising from Eq. 6.68, suppose $N_x = 3$ and $N_y = 4$ as shown in Fig. 6.4.

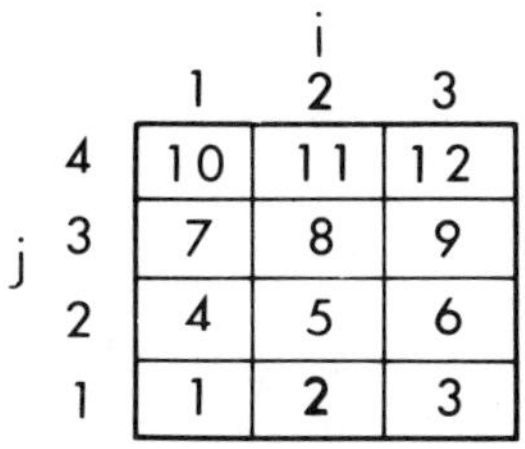

	i		
	1	2	3
j=4	10	11	12
j=3	7	8	9
j=2	4	5	6
j=1	1	2	3

Fig. 6.4 Rectangular 2-D Reservoir.

The numbers around the periphery of the larger rectangle are the (i,j) coordinate indices. The grid blocks are sequentially numbered beginning at the bottom progressing upward from left to right. We call these numbers the *block indices* and the ordering *normal grid ordering*. The block indices, m, are related to the coordinate indices by the relationship $m = i + (j-1)N_x$. For example, if $i = 2$ and $j = 2$ then $m = 5$. Notice also that the left-hand side of Eq. 6.68 involves the blocks immediately to the right and left and above and below the (i,j) block. If the block index corresponding to the (i,j) block is m, then the blocks to the left, right, above and below will have indices $m-1$, $m+1$, $m+N_x$, $m-N_x$, respectively. This is readily seen by referring to Fig. 6.4. Consequently, we can express Eq. 6.68 in terms of the single block index, i.e.,

$$B_m p_{m-N_x} + D_m p_{m-1} + E_m p_m + F_m p_{m+1} + H_m p_{m+N_x} = \tilde{q}_m,$$
$$m = 1, 2, \ldots, N \quad (6.69)$$

where $N = N_x N_y$, $B_m = -(T_y)_{j-1/2}$, $D_m = -(T_x)_{i-1/2}$, $E_m = \Sigma T$, $F_m = -(T_x)_{i+1/2}$, and $H_m = -(T_y)_{j+1/2}$. When m takes on values equal to the block indices on the boundary, then some of the terms in Eq. 6.69 refer to pressures in blocks not included in the system. In these cases, the boundary conditions will be such that these terms can be deleted from the system of equations. In fact, the coefficients multiplying those terms will usually be zero. For purposes of illustration, we assume the latter is true. Consequently, for the 3×4 problem, Eq. 6.69 generates the following system of equations:

$$
\begin{aligned}
E_1 p_1 + F_1 p_2 \quad\quad\quad + H_1 p_4 \quad\quad\quad\quad\quad\quad\quad\quad\quad\quad\quad\quad\quad\quad\quad\quad\quad\quad &= \bar{q}_1 \\
D_2 p_1 + E_2 p_2 + F_2 p_3 \quad\quad + H_2 p_5 \quad\quad\quad\quad\quad\quad\quad\quad\quad\quad\quad\quad\quad\quad\quad &= \bar{q}_2 \\
\quad\quad + D_3 p_2 + E_3 p_3 \quad\quad\quad\quad + H_3 p_6 \quad\quad\quad\quad\quad\quad\quad\quad\quad\quad\quad\quad &= \bar{q}_3 \\
B_4 p_1 \quad\quad\quad\quad\quad\quad + E_4 p_4 + F_4 p_5 \quad\quad + H_4 p_7 \quad\quad\quad\quad\quad\quad\quad\quad\quad\quad &= \bar{q}_4 \\
\quad\quad B_5 p_2 \quad\quad\quad\quad + D_5 p_4 + E_5 p_5 + F_5 p_6 \quad\quad + H_5 p_8 \quad\quad\quad\quad\quad\quad\quad &= \bar{q}_5 \\
\quad\quad\quad\quad B_6 p_3 \quad\quad\quad\quad + D_6 p_5 + E_6 p_6 \quad\quad + H_6 p_9 \quad\quad\quad\quad\quad\quad &= \bar{q}_6 \\
\quad\quad\quad\quad\quad\quad B_7 p_4 \quad\quad\quad\quad + E_7 p_7 + F_7 p_8 \quad\quad + H_7 p_{10} \quad\quad\quad\quad &= \bar{q}_7 \\
\quad\quad\quad\quad\quad\quad\quad\quad B_8 p_5 \quad\quad + D_8 p_7 + E_8 p_8 + F_8 p_9 \quad\quad + H_8 p_{11} \quad\quad &= \bar{q}_8 \\
\quad\quad\quad\quad\quad\quad\quad\quad\quad\quad B_9 p_6 \quad\quad + D_9 p_8 + E_9 p_9 \quad\quad + H_9 p_{12} &= \bar{q}_9 \\
\quad\quad\quad\quad\quad\quad\quad\quad\quad\quad\quad\quad B_{10} p_7 \quad\quad\quad\quad + E_{10} p_{10} + F_{10} p_{11} \quad\quad &= \bar{q}_{10} \\
\quad\quad\quad\quad\quad\quad\quad\quad\quad\quad\quad\quad\quad\quad B_{11} p_8 \quad\quad + D_{11} p_{10} + E_{11} p_{11} + F_{11} p_{12} &= \bar{q}_{11} \\
\quad\quad\quad\quad\quad\quad\quad\quad\quad\quad\quad\quad\quad\quad\quad\quad B_{12} p_9 \quad\quad + D_{12} p_{11} + E_{12} p_{12} &= \bar{q}_{12}
\end{aligned}
$$
$$\dots \quad (6.70)$$

In matrix form this becomes $\mathbf{Ap} = \mathbf{q}$ where $\mathbf{A}$ is of the form shown in Fig. 6.5(a). If this had been a 3-D problem, the structure would be the one shown in Fig. 6.5(b). The banded structures give rise to the name *band matrices*. The nonzero elements occur only on the solid parallel diagonals shown in Figs. 6.5(a) and 6.5(b). All other elements are zero. If the order of the matrix is large, the ratio of zeros to nonzeros is high, and the matrix is said to be *sparse*. Those one encounters in reservoir simulation are frequently sparse. Note in Figs. 6.5(a) and 6.5(b) that all the elements outside

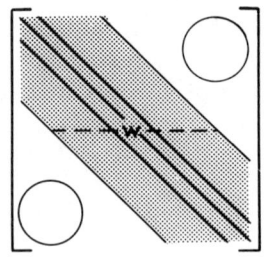
A. Pentadiagonal Matrix Structure

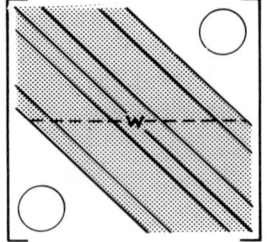
B. Heptadiagonal Matrix Structure

Fig. 6.5 Matrix Structures for 2-D and 3-D Problems.

of a band of width w centered on the main diagonal are zero. For our example 3×4 problem, the band width is $w = 7$. For normal grid ordering, the band width is equal to $2N_x + 1$ in 2-D and $2N_xN_y + 1$ in 3-D (taking the z-direction last). The amount of work involved in treating band matrix problems is roughly proportional to the square of the band width. Consequently, N_x and N_y should correspond to the shortest sides to minimize the band width.

6.8.1 Determination of Matrix Structures

To determine the matrix structure for a reservoir problem, we need only know the dimensionality of the reservoir. Consider, for example, a

Column Numbers

Row Numbers	1	2	3	4	5
1	x	x			
2	x	x	x		
3		x	x	x	
4			x	x	x
5				x	x

Fig. 6.6 Matrix Structure Setup.

1-D reservoir problem consisting of a collection of contiguous grid blocks. Suppose there are 5 blocks numbered 1 through 5. The matrix structure for flow through this reservoir is obtained as follows: Using a piece of gridded paper (graph paper is satisfactory), number 5 columns across the sheet and 5 rows down such that the numbers are equally spaced. This defines 25 row-column locations in all as depicted in Fig. 6.6.

Each grid block in the reservoir represents a unique row in the matrix. Furthermore, flow can potentially occur in each block to one or more blocks, through the faces between them. This determines the column locations where the nonzero elements occur in each row. For example, block 1 can support flow, but only to block 2. Thus for the first row in Fig. 6.6, we insert x's denoting nonzero elements in the first and second columns. Similarly, block 2 supports flow, possibly to 1 or to 3. Thus, in the second row above, we insert x's in columns 1, 2 and 3. This procedure is continued for all 5 blocks, and a tridiagonal structure emerges. Extending this idea to the 2-D problem in Fig. 6.4 gives the structure shown in Fig. 6.7.

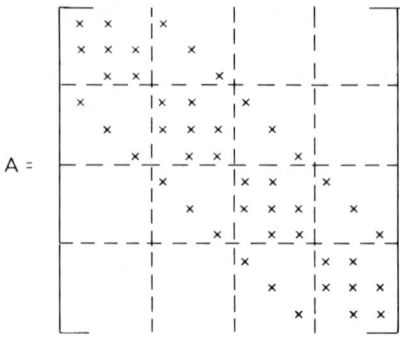

Fig. 6.7 Matrix Structure for a 3 × 4 × 1 Areal Reservoir Problem.

For single-phase flow, the off-diagonal elements in the matrix consist of transmissibilities at the block interfaces. Each diagonal element consists of the sums of the transmissibilities about the faces of the grid block corresponding to the row, plus some possible additional terms. The additional terms come from a well within the block (if any) and terms brought over from the right-hand sides of the finite difference equations. The net effect is that the matrix will usually be strictly diagonally dominant. (See Appendix A.6.9).

A direct solution technique for band matrices can be obtained by employing **LU** decomposition (see Appendix A.6.10). This procedure involves three steps: (1) factorization, (2) forward solution, (3) backward solution. Define $\hat{w} = (w-1)/2$ and let

$$L_1(i) \equiv \max\,(1,\, i-\hat{w})$$

$$L_2(i) \equiv \min\,(N,\, i+\hat{w})$$

$$L_3(j) \equiv \max\,(1,\, j-\hat{w})$$

The Crout algorithm for **LU** decomposition of a band matrix is

$$l_{ij} = a_{ij} - \sum_{k=L_1}^{j-1} l_{ik} u_{kj},\; j = L_1,\, L_1+1,\, \ldots,\, i \qquad (6.71)$$

$$u_{ij} = (a_{ij} - \sum_{k=L_3}^{i-1} l_{ik} u_{kj})/l_{ii},\; j = i+1,\, \ldots,\, L_2(i) \qquad (6.72)$$

$$y_i = (b_i - \sum_{k=L_1}^{i-1} l_{ik} y_k)/l_{ii},\; i = 1,\, 2\, \ldots,\, N \qquad (6.73)$$

$$x_i = y_i - \sum_{k=i+1}^{L_2} u_{ik} x_k, \; i = N, \, N-1, \ldots, 1 \tag{6.74}$$

Eqs. 6.71 and 6.72 are the factorization steps, Eq. 6.73 is the forward solution and Eq. 6.74 is the back solution.

This algorithm takes advantage of the banded structure of the matrix since it eliminates operations on the zero elements outside the band. However, there are still some wasted operations on the zeros within the band. By wasted operations, we mean addition or multiplication by zero. When the bandwidth is large, the algorithm becomes less efficient. In such cases, special sparse matrix techniques can be employed which involve operations only on the nonzero elements. If there are no zeros in the band, as in a tridiagonal matrix, the band algorithm cannot be surpassed in efficiency. Indeed, the Thomas tridiagonal algorithm cited in chapter 5 is a special case of the band algorithm and can be easily derived from it.

6.9 Ordering Schemes and Sparse Matrix Methods

Recall that in Gaussian elimination, we create an upper triangular matrix **U** from the original matrix **A**. This involves eliminating nonzeros below the main diagonal, and processing rows 2, 3, ..., N one at a time. In so doing, we sometimes create nonzeros, below the current row being processed, where zero elements originally existed. These are called "fill" elements which must be stored and eventually eliminated themselves. By reordering the equations, we sometimes can reduce both storage and work requirements. This is accomplished by exploiting the matrix sparsity and reducing the number of fill elements created. For example, suppose we have a matrix with the structure depicted in Fig. 6.8 where x represents a nonzero element. If we upper triangulate this matrix as if it was full, we perform $(n^3 + 3n^2 - n)/3 = 36$ arithmetic operations (divides and multiplies). However, observe that all elements below the main diagonal, with the exception of a_{41}, are already zero. Thus, if we skip to the last equation and eliminate a_{41} only, we achieve the upper triangularization in 12 operations. Furthermore, this is accomplished without creating fill. Essentially, this amounts to reordering the rows of the matrix. For example, if the computer code recognizes that once the last row is processed, the elimination is complete, and it processes the last row first, it will stop thereafter, avoiding unnecessary work.

Obviously, developing computer code that can look ahead, assess where the nonzeros and potential fill elements are located, and then arrive at

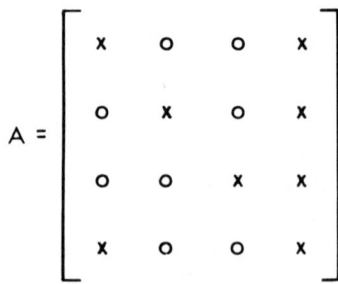

Fig. 6.8 Hypothetical Matrix Structure.

the best reordering to minimize work and storage is rather tricky. However, it can be done. Software packages that do this are called *sparse matrix routines*.[15] Their common characteristics are reordering of the matrix elements, followed by operations on the nonzeros only. Such routines frequently employ LU decomposition for solving the matrix problem, rather than Gaussian elimination, because the former procedure preserves the original sparsity of the matrix as much as possible where Gaussian elimination does not.[16]

Reordering the matrix elements is equivalent to renumbering the grid blocks of the reservoir. For example, consider a reservoir that is shaded checkerboard fashion as in Fig. 6.9. If we number the dark blocks first and then the white blocks, from top to bottom, beginning from the left, we get a substantially different matrix structure. A typical matrix is shown for alternate diagonal ordering of a 30 × 30 system in Fig. 6.10. This is also called D-4 or white-black ordering. This ordering scheme was first employed in the electrical power industry for network problems[17] and subsequently discussed by Price and Coats in the petroleum literature.[18] Note that it's necessary to triangularize only the lower half of the matrix. Conse-

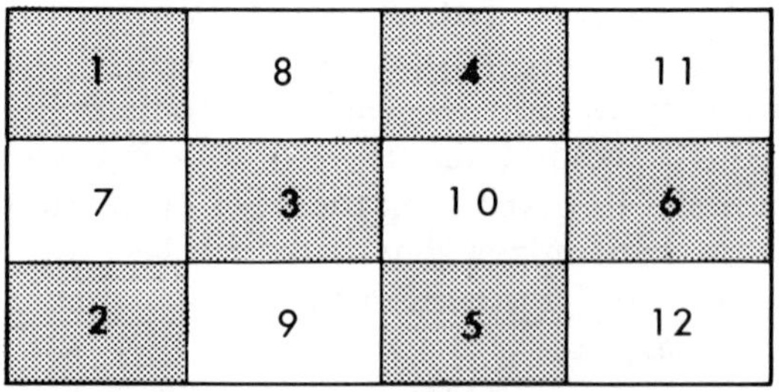

Fig. 6.9 Alternate Diagonal Ordering.

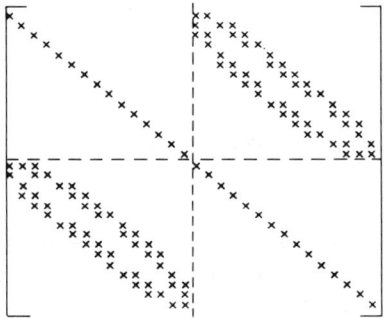

Fig. 6.10 Matrix Structure for Alternate Diagonal Ordering.

quently, the work† and storage is cut by a factor of at least two during Gaussian elimination. When the triangularization is completed, the matrix U has the form shown in Fig. 6.11 where the small circles are fill elements.

Some simulators offer the option of using alternate diagonal ordering in conjunction with Gaussian elimination. For certain problems, this can

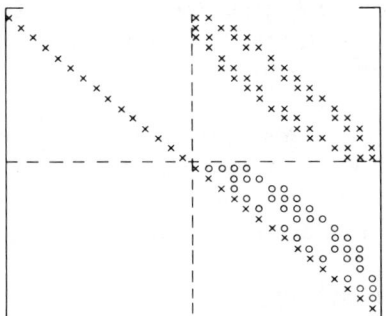

Fig. 6.11 Matrix U from Alternate Diagonal Ordering.

produce substantial savings in cost and computer storage. For example, for a 2-D areal problem we have $W_1 \simeq IJ^3$ and $S_1 \simeq IJ^2$ for large I and J ($J < I$, $I = N_x$, $J = N_y$), where W_1 and S_1 are work and storage, respectively for normal grid ordering. For D-4 ordering,

$$W_4 \simeq IJ^3/2 - J^4/4 \tag{6.75}$$

$$S_4 \simeq IJ^2/2 - J^3/6 \tag{6.76}$$

---

† We use "work" in the sense of operations a computer must perform. This includes adds, multiplies, divides, fetches and stores.

For 2-D problems,

$$W_4/W_1 = (2\ell - 1)/4\ell \qquad (6.77)$$

where $\ell = I/J$. Thus, for highly elongated rectangles $(I > J)$, D-4 is twice as fast as standard Gaussian elimination and four times faster when $\ell = 1$.

In Appendix A.6.8, we provide some basic definitions regarding the graph of a matrix, **A**. The notation and theory of graphs is especially useful in treating sparse matrices.[19] With regard to the latter, our concern is primarily with undirected graphs. It would be nice if one could find an ordering that produced a matrix whose graph is a tree since, in this case, Gaussian elimination produces no fill. However, for the banded matrices encountered in reservoir simulation, this does not appear possible. The best one can hope for is an optimal or near-optimal ordering that minimizes the number of fill elements. To date such approaches have not been fully investigated relative to reservoir simulation problems. A number of near-optimal ordering algorithms have been proposed for electrical networks.[20] They are not guaranteed, however, to produce the minimum in fills, but in many practical cases, they do so.

Given a matrix problem $\mathbf{A}x = \mathbf{b}$, elimination of unknowns is equivalent to eliminating nodes in the graph G of **A** to produce a subgraph G'. The subgraph is derived from G as follows:

(1) For every pair of edges (a,b) and (a,c) in G incident at node a, construct a new edge joining b and c. If this edge is connected in parallel with an existing one, represent the two as a single edge.
(2) Delete node a and all edges incident at node a. The resulting graph is G'.
(3) If in step 1 the new edge is not a parallel edge with an existing one, then a fill element will be created in the upper triangular matrix **U**.†

To illustrate, consider the 6×6 matrix in Fig. 6.12 where, again, the x's represent nonzeros. The graph of **A** is displayed in Fig. 6.13 where we omit the loops corresponding to the diagonal elements. If the elimination of unknowns is in the order, $x_1, x_2, \ldots$, then we get the subgraphs shown in Fig. 6.14 for the first four eliminations. Obviously no fill will occur beyond this point. The total fill is 4 (2 each in the **L** and **U** matrices). The idea in optimal ordering schemes from the graph theory point of view, is to seek

---

† The **U**-matrix created from **LU**-decomposition is identical with that obtained by Gaussian elimination. Consequently the fill in **U** will be the same for both methods. Therefore, it suffices to determine the fill in **U** only, and double it to get the fill in **L** and **U**.

Single-Phase Multidimensional Flow

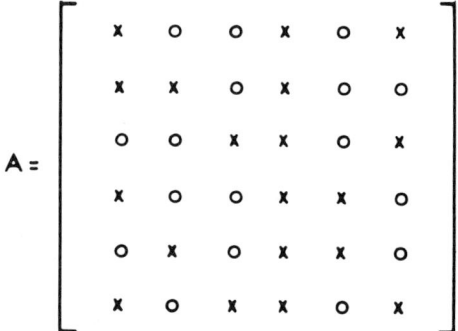

Fig. 6.12 Example 6 × 6 Matrix.

that ordering that produces a collection of subgraphs with the minimum number of new edges.

A typical near-optimal ordering scheme is Berry's.[21] He describes it in terms of matrix manipulations. Here we describe it in terms of equivalent matrix graph reductions of an nth order matrix. Assume the *incidence matrix*

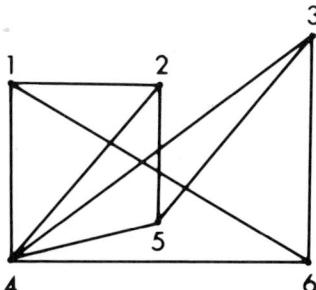

Fig. 6.13 Graph of Matrix A.

of A is symmetric,† $a_{ii} \neq 0$ for all i, and when LU factorization is performed, $l_{ii} \neq 0$ for all i. The algorithm is presented in flow chart form in Fig. 6.15. If we apply this to the graph in Fig. 6.13, we find it fails the first two tests; i.e., there are no nodes with only one incident edge, nor are there any, which when removed, yield a parallel edge with an existing edge. Consequently, we must seek a node yielding minimum fill. Removal of any one of nodes 1, 2, 3, 5, or 6 will yield one fill element (a tie). Removal

---

† The incidence matrix, M say, of A is one where $m_{ij} = 1$ if $a_{ij} \neq 0$ and $m_{ij} = 0$ if $a_{ij} = 0$.

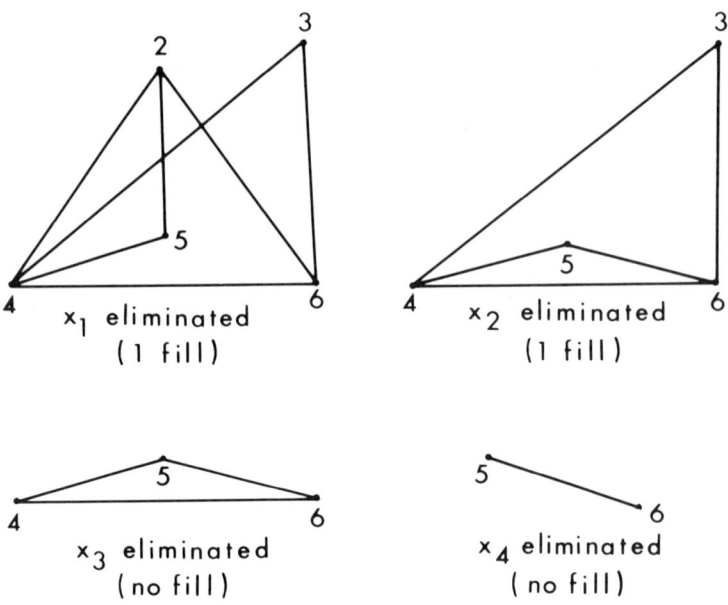

Fig. 6.14 Subgraphs of Matrix A.

of node 4 first (the worst choice) yields five fills. Each of nodes 1, 2, 3, 5, and 6 have three incident edges so it is immaterial which of these is selected as the first eliminant. Suppose node 1 is selected. The resultant subgraph is in the upper left-hand corner of Fig. 6.14. The first test in the flow chart is again applied and the subgraph fails; however, it does not fail the second test. Clearly, node 5 upon removal yields a parallel edge, hence it is renumbered as 2. Finally nodes 2, 3, and 4 are renumbered and removed. The entire process results in two fill elements after doubling. The renumbering is as follows: $1 \rightarrow ①$; $5 \rightarrow ②$; $2 \rightarrow ③$; $3 \rightarrow ④$; $4 \rightarrow ⑤$; and $6 \rightarrow ⑥$ where the circled values represent the new node orderings. Notice that $1 \rightarrow ①$; $5 \rightarrow ②$; $3 \rightarrow ③$; $2 \rightarrow ④$; $6 \rightarrow ⑤$; $4 \rightarrow ⑥$ also yields two fill elements.

6.10 Strongly Implicit Procedure

For reservoir problems requiring a large number of grid blocks, the storage requirements can become excessively high for direct solution methods. Even sparse matrix techniques can lose their utility since a substantial overhead is incurred in maintaining pointers for the original and created nonzeros.[15] Moreover, all advantages of sparse matrix routines are lost if they are programmed poorly. Under such circumstances, an alternative is to employ

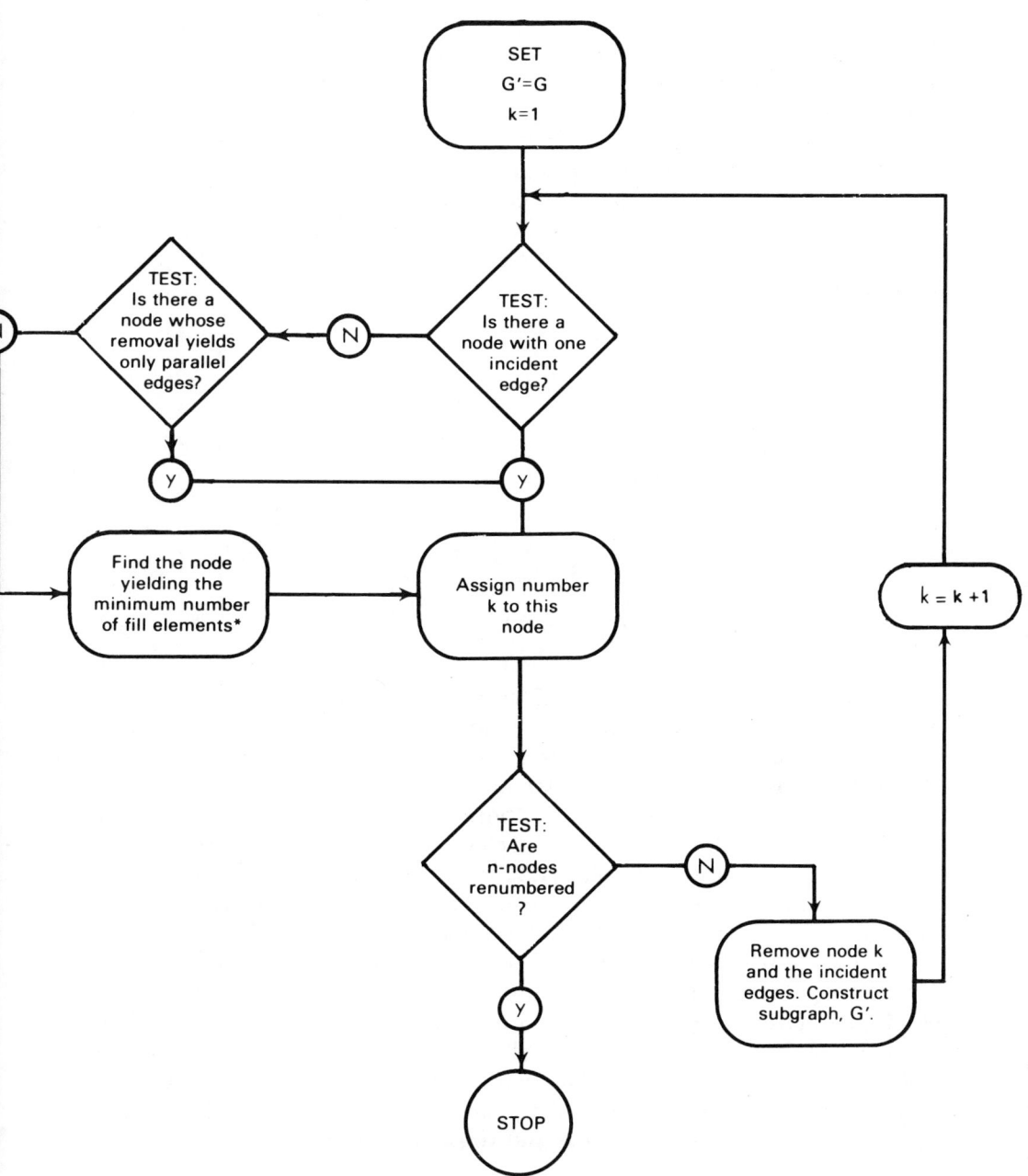

* To achieve this, each node is examined and the fills are determined assuming that particular node only is eliminated. The node leading to the minimum number of fills is selected. For ties, the node with the maximum number of incident edges is selected.

Fig. 6.15 Berry's Near-Optimal Ordering Algorithm.[21]

iterative techniques. Their principal advantage is that storage requirements are minimal. The main disadvantage is that some iterative techniques, under the best of circumstances, fail to converge. This occurs frequently with the IADI methods where the transmissibilities have a high contrast in one direction relative to the others. As an alternate iterative method Stone[22] introduced the Strongly Implicit Procedure (SIP) that readily handles many of the difficult problems of reservoir simulation. Furthermore, it has the advantage that iteration parameters can be more easily selected. It is equally applicable to parabolic and elliptic systems in both 2-D and 3-D. We briefly discuss the SIP algorithm here in the context of the 2-D parabolic problem given in Eq. 6.12.

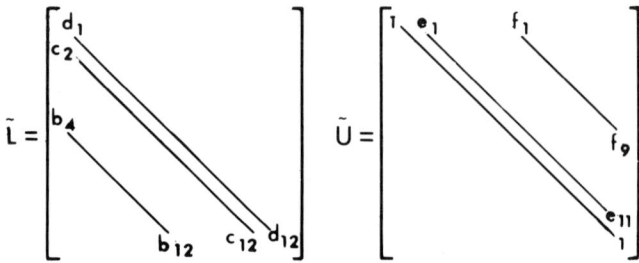

Fig. 6.16 $\tilde{L}$ and $\tilde{U}$ for a 3×4 problem.

Consider the problem $Ax = b$. The idea is to replace A with a matrix $\tilde{A}$ that is sufficiently close to A in some sense, but yields a faster rate of convergence when used as an iteration matrix. Suppose we construct $\tilde{A}$ and factor it into a lower triangular matrix $\tilde{L}$, and a unit upper triangular matrix $\tilde{U}$. In each of these factors there are at most three elements per row (for a 2-D problem). Since $\tilde{A}$ is close to A, factorization of $\tilde{A}$ can be regarded as an approximate factorization of A. The $\tilde{L}$ and $\tilde{U}$ matrices are then used in a sequential forward and backward solution procedure. In this regard, the steps involved, i.e., factorization, forward solution, and back solution are essentially those in LU decomposition described previously.

The algebraic expansion of Eq. 6.12 is given in Eq. 6.69. In Fig. 6.16 we depict factors $\tilde{L}$ and $\tilde{U}$ for the 3×4 problem considered in section 8. If we premultiply $\tilde{U}$ by $\tilde{L}$ then $\tilde{A}$ has the structure shown in Fig. 6.17. Compare A and $\tilde{A}$ and note that two additional diagonals are introduced that fall just inside the $\tilde{B}$ and $\tilde{H}$ diagonals. We have the following relationships between the elements in $\tilde{A}$ and those in $\tilde{L}$ and $\tilde{U}$.

$$\tilde{B}_m = b_m \tag{6.78a}$$

$$\tilde{C}_m = b_m e_{m-N_x} \tag{6.78b}$$

$$\tilde{D}_m = c_m \tag{6.78c}$$

$$\tilde{E}_m = b_m f_{m-N_x} + c_m e_{m-1} + d_m \tag{6.78d}$$

$$\tilde{F}_m = d_m e_m \tag{6.78e}$$

$$\tilde{G}_m = c_m f_{m-1} \tag{6.78f}$$

$$\tilde{H}_m = d_m f_m \tag{6.78g}$$

These relationships must all be satisfied so that $\tilde{A}$ is factorable. On the other hand, we have for every block m, the five unknowns b_m, c_m, d_m, e_m,

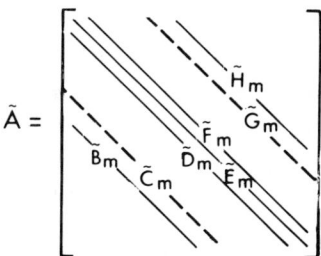

Fig. 6.17 Structure of $\tilde{A}$.

f_m. Thus, of the seven relationships in Eq. 6.78 five of the seven capitalized coefficients are arbitrary while the remaining two can be related to the other five using Eq. 6.78. Of course it is also necessary to relate the coefficients of $\tilde{A}$ to A. Stone made the following choices:

$$\tilde{B}_m = B_m - \sigma \tilde{C}_m \tag{6.79a}$$

$$\tilde{D}_m = D_m - \sigma \tilde{G}_m \tag{6.79b}$$

$$\tilde{E}_m = E_m + \sigma(\tilde{C}_m + \tilde{G}_m) \tag{6.79c}$$

$$\tilde{F}_m = F_m - \sigma \tilde{C}_m \tag{6.79d}$$

$$\tilde{H}_m = H_m - \sigma \tilde{G}_m \tag{6.79e}$$

where σ is an iteration parameter. These relationships lead to the matrix equation

$$B_m p_{m-N_x} + \tilde{C}_m \bar{p} + D_m p_{m-1} + E_m p_m + F_m p_{m+1}$$
$$+ \tilde{G}_m \hat{p} + H_m p_{m+N_x} = \tilde{q}_m \qquad (6.80)$$

where

$$\bar{p} = p_{m+1-N_x} - \sigma(p_{m+1} + p_{m-N_x} - p_m), \qquad (6.81)$$

$$\hat{p} = p_{m-1+N_x} - \sigma(p_{m-1} + p_{m+N_x} - p_m). \qquad (6.82)$$

Note that Eq. 6.80 is identical with Eq. 6.69 except for the additional terms $\tilde{C}_m \bar{p}$ and $\tilde{G}_m \hat{p}$ which introduce the additional points at $m + 1 - N_x$ and $m - 1 + N_x$. If $\bar{p}$ and $\hat{p}$ are sufficiently small, $\tilde{A}$ will be close to A in some sense. This is especially true if the pressures are smooth over the grid. Note also that the computational star is now

```
                •            •
            m − 1 + N_x    m + N_x
                •            •           •
              m − 1          m         m + 1
                             •           •
                           m − N_x   m + 1 − N_x.
```

If we combine Eqs. 6.79 and 6.78, we get the SIP factorization algorithm as follows:

$$b_m = \frac{B_m}{1 + \sigma e_{m-N_x}} \qquad (6.83)$$

$$c_m = \frac{D_m}{1 + \sigma f_{m-1}} \qquad (6.84)$$

$$d_m = E_m + \sigma(b_m e_{m-N_x} + c_m f_{m-1}) - b_m f_{m-N_x} - c_m e_{m-1} \qquad (6.85)$$

$$e_m = \frac{F_m - \sigma b_m e_{m-N_x}}{d_m} \qquad (6.86)$$

$$f_m = \frac{H_m - \sigma c_m f_{m-1}}{d_m} \qquad (6.87)$$

These five equations can be solved sequentially for $m = 1, 2, \ldots, N$.

Once the factorization is completed using Eqs. 6.83–6.87, we use $\tilde{A}$ as an iteration matrix, i.e., for the problem $\tilde{A}p = q$, write

$$\tilde{A}p^{k+1} = \tilde{A}p^k - \omega(Ap^k - q). \tag{6.88}$$

Define a displacement vector DP^{k+1} and residual vector R^k by

$$DP^{k+1} \equiv p^{k+1} - p^k$$

$$R^k \equiv q^k - Ap^k$$

such that Eq. 6.88 becomes

$$\tilde{A}DP^{k+1} = \omega R^k \tag{6.89}$$

with ω as an iteration parameter. Now Eq. 6.89 can be written as

$$\tilde{L}\tilde{U}DP^{k+1} = \omega R^k$$

or

$$\tilde{L}v^{k+1} = \omega R^k, \tag{6.90}$$

and

$$\tilde{U}DP^{k+1} = v^{k+1}. \tag{6.91}$$

Eqs. 6.90 and 6.91 are the forward and back solutions, respectively. Expanding each yields

$$b_m v_{m-N_x}^{k+1} + c_m v_{m-1}^{k+1} + d_m v_m^{k+1} = \omega R_m^k \tag{6.92}$$

$$DP_m^{k+1} + e_m DP_{m+1}^{k+1} + f_m DP_{m+N_x}^{k+1} = v_m^{k+1} \tag{6.93}$$

Eq. 6.92 is solved for $m = 1, 2, \ldots, N$ while in Eq. 6.93 $m = N - 1$, $N - 2, \ldots, 1$.

An alternative procedure to Stone's was proposed by Dupont, Kendall, and Rachford[23] where in Eqs. 6.81 and 6.82 they let $\tilde{p} = p_{m+1-N_x} - p_m$ and $\hat{p} = p_{m-1+N_x} - p_m$ to reduce the effects of $\tilde{C}_m$ and $\tilde{G}_m$. In addition, they replace E_m by $E_m + \beta$ where the latter is an iteration parameter. However, Stone's method appears to be superior. In the DKR method, varying ω can help to accelerate convergence, however, in Stone's method a fixed value of one is usually satisfactory. It is important to cycle through a set of σ-values. The minimum value is not critical and can be set to zero. The maximum value is the most sensitive parameter and can be estimated from $1 - \sigma_1$ where σ_1 is given in Eq. 6.24. As in ADI the parameters are geometrically spaced. For cases where the transmissibility ratios in Eq. 6.24 vary over the grid, local values of σ_{max} are computed and the maximum over the grid is used.

Although the algorithm presented here is for a 2-D single-phase flow problem, the technique has been extended by Weinstein, Stone, and Kwan[24] to 3-D problems and problems involving multiphase flow[25]. In all cases, except for 2-D single-phase flow in isotropic, homogeneous media, SIP achieved greater rates of convergence than iterative ADI. As the degree of the nonlinearities increases, the advantage of SIP over ADI appears to increase, and in many instances, no difficulty with SIP is experienced when ADI fails to converge at all.

6.11 Point Iterative Methods

For a given matrix problem, $\mathbf{Ax} = \mathbf{b}$, the matrix $\mathbf{A}$ can be expressed as the sum $\mathbf{A} = \mathbf{D} - \mathbf{L} - \mathbf{U}$. $\mathbf{D}$ is a diagonal matrix containing the diagonal elements of $\mathbf{A}$, and $\mathbf{L}$ and $\mathbf{U}$ are strict lower and upper triangular matrices whose elements are the negatives of the elements of $\mathbf{A}$ below and above the main diagonal of $\mathbf{A}$. We write the problem as

$$\mathbf{Dx} = (\mathbf{L} + \mathbf{U})\mathbf{x} + \mathbf{b} \qquad (6.94)$$

$$\mathbf{x}^{(k+1)} = \mathbf{M}_J \mathbf{x}^{(k)} + \mathbf{c} \qquad (6.95)$$

where $\mathbf{M}_J \equiv \mathbf{D}^{-1}(\mathbf{L} + \mathbf{U})$, the *point Jacobi* iteration matrix and $\mathbf{c} \equiv \mathbf{D}^{-1}\mathbf{b}$. This procedure requires that one store all the components of the vector $\mathbf{x}^{(k)}$ while computing $\mathbf{x}^{(k+1)}$. An approach that does not require this is the *point Gauss-Seidel* method where the iteration matrix is $\mathbf{M}_{GS} \equiv (\mathbf{D} - \mathbf{L})^{-1}\mathbf{U}$ and $\mathbf{c} = (\mathbf{D} - \mathbf{L})^{-1}\mathbf{b}$. In reservoir simulation neither of these techniques are employed since the convergence rates are too slow. However, convergence can be accelerated using an amplification factor (or iteration parameter), ω. We define the displacement vector $\mathbf{d} = \mathbf{x}^{(k+1)} - \mathbf{x}^{(k)}$, i.e., it is the change in the solution vector over one iteration. The *point successive overrelaxation* (SOR) method is based on amplifying the displacement vectors obtained from the Gauss-Seidel procedure, i.e.,

$$\mathbf{x}^{(k+1)} = \mathbf{x}^{(k)} + \omega \mathbf{d}_{GS}. \qquad (6.96)$$

This leads to the following matrix form:

$$\mathbf{x}^{(k+1)} = (\mathbf{D} - \omega \mathbf{L})^{-1} \{(1 - \omega)\mathbf{D} + \omega \mathbf{U}\} \mathbf{x}^{(k)} + \omega(\mathbf{D} - \omega \mathbf{L})^{-1}\mathbf{b} \qquad (6.97)$$

where the iteration matrix is $\mathbf{M}_\omega \equiv (\mathbf{D} - \omega\mathbf{L})^{-1}\{(1 - \omega)\mathbf{D} + \omega\mathbf{U}\}$ and the vector $\mathbf{c}$ is defined by the rightmost term in Eq. 6.97.

These techniques are referred to as "point" methods since the components in vector $\mathbf{x}$ are computed sequentially point-by-point. For example, the SOR method in terms of individual components becomes

$$a_{ii}x_i^{(k+1)} = a_{ii}(1-\omega)x_i^{(k)} - \omega\left\{\sum_{j=1}^{i-1} a_{ij}x_j^{(k+1)} + \sum_{j=i+1}^{n} a_{ij}x_j^{(k)} - b_i\right\} \quad (6.98)$$

for an nth order matrix problem. A necessary and sufficient condition for convergence of the three techniques cited here is that the spectral radius of the iteration matrix be less than one.[26] Obviously, the spectral radius in the SOR method is determined by the choice of ω. It has been found that the SOR method will converge when $0 < \omega < 2$ and whenever $\mathbf{A}$ is a positive definite matrix.[27] This is frequently the case with reservoir simulation problems. The fastest convergence rate with SOR is obtained when an optimum iteration parameter, ω_o, is employed. By optimum we mean that choice of ω that produces the smallest possible value of $\rho(\mathbf{M}_\omega)$. An examination of convergence rates indicates that $1 < \omega_o < 2$, consequently, the choice of ω should be made within these bounds. Unfortunately, there is no easy way to compute ω_o for practical reservoir simulation problems. For single-phase flow in a $1 - D$, homogeneous, isotropic reservoir ω_o is given by

$$\omega_o = \frac{2}{1 + \sqrt{1-\sigma^2}} \quad (6.99)$$

where $\sigma \equiv \rho(\mathbf{M}_J)$. This can possibly be used for an initial estimate for multiphase, heterogeneous, anisotropic systems. Subsequent numerical experimentation with the simulator can then produce a better estimate. As an alternative to this, several mathematical procedures[28-31] are available for estimating ω_o.

Note that when $\omega = 1$, SOR becomes the Gauss-Seidel method. The convergence rate of the latter is approximately twice that of the Jacobi technique, while for a proper choice of ω, SOR is capable of an even faster rate.

6.12 Block Successive Overrelaxation

A point SOR method is readily extended to a *block* SOR technique by considering a partitioning of the problem $\mathbf{Ax} = \mathbf{b}$, e.g.,

$$\begin{bmatrix} A_{11} & A_{12} & \cdots & A_{1n} \\ A_{21} & A_{22} & \cdots & A_{2n} \\ \vdots & \vdots & & \vdots \\ A_{n1} & A_{n2} & \cdots & A_{nn} \end{bmatrix} \begin{bmatrix} \mathbf{x}_1 \\ \mathbf{x}_2 \\ \vdots \\ \mathbf{x}_n \end{bmatrix} = \begin{bmatrix} \mathbf{b}_1 \\ \mathbf{b}_2 \\ \vdots \\ \mathbf{b}_n \end{bmatrix} \qquad (6.100)$$

where the A_{ij}'s are submatrices and the $\mathbf{x}_i$'s and $\mathbf{b}_i$'s contain subcomponents of $\mathbf{x}$ and $\mathbf{b}$. For example, the dashed lines on the pentadiagonal matrix shown in Fig. 6.7 define a partitioning into *block tridiagonal* form. The diagonal submatrices A_{ii} are themselves tridiagonal, while the off-diagonal submatrices are diagonal matrices. In this case, the partitions of the vectors $\mathbf{x}$ and $\mathbf{b}$ would each contain three components: $\mathbf{x}_1 = (x_1, x_2, x_3)^T$, $\mathbf{b}_1 = (b_1, b_2, b_3)^T$, $\mathbf{x}_2 = (x_4, x_5, x_6)^T$, $\mathbf{b}_2 = (b_4, b_5, b_6)^T$, etc. A block SOR technique utilizes the partitioned matrix form and replaces the scalar calculations in Eq. 6.98 with matrix calculations. Thus we have the analog of Eq. 6.98,

$$A_{ii}\mathbf{x}_i^{(k+1)} = A_{ii}(1-\omega)\mathbf{x}_i^{(k)} - \omega \left\{ \sum_{j=1}^{i-1} A_{ij}\mathbf{x}_j^{(k+1)} + \sum_{j=i+1}^{n} A_{ij}\mathbf{x}_j^{(k)} - \mathbf{b}_i \right\}$$
$$i = 1, \ldots, n. \quad (6.101)$$

A better computational algorithm is

$$A_{ii}\mathbf{y}_i^{(k+1)} = -\sum_{j=1}^{i-1} A_{ij}\mathbf{x}_j^{(k+1)} - \sum_{j=i+1}^{n} A_{ij}\mathbf{x}_j^{(k)} + \mathbf{b}_i, \; i = 1, 2, \ldots, n \qquad (6.102)$$

$$\mathbf{x}_i^{(k+1)} = \omega \left\{ \mathbf{y}_i^{(k+1)} - \mathbf{x}_i^{(k)} \right\} + \mathbf{x}_i^{(k)}, \; i = 1, 2, \ldots, n \qquad (6.103)$$

where $\mathbf{y}_i^{(k+1)}$ is an intermediate vector. For the matrix displayed in Fig. 6.7, the intermediate vector can be obtained by the Thomas tridiagonal algorithm. Observe that each application of Eqs. 6.102 and 6.103 for $i = 1, 2, 3$ and 4 yields one line of unknowns for the rectangular reservoir in Fig. 6.4. Thus this approach is called *line successive overrelaxation* (LSOR). If we partition the matrix about the central lines of symmetry in Fig. 6.7 such that $\mathbf{A} = \begin{bmatrix} A_{11} & A_{12} \\ A_{21} & A_{22} \end{bmatrix}$ then the diagonal submatrices are pentadiagonal. The algorithm in Eqs. 6.102–6.103 is still applicable with the exception that the matrix problem in Eq. 6.102 is solved by the band algorithm (or possibly a sparse matrix technique for larger problems). Since this partitioning yields a simultaneous solution for the unknowns in two rows of grid blocks, it is called *two-line successive overrelaxation* (2LSOR).

The reason for passing from point to block SOR methods is that convergence rates are significantly better with the latter. Moreover, block methods generally incur less round-off error and can be performed in such a way

Single-Phase Multidimensional Flow

that the number of arithmetic operations are essentially the same as the point SOR method. For these reasons, point SOR is rarely, if ever, used in commercial simulators. In fact, of the iterative techniques discussed in this chapter, SIP and block SOR are most frequently used.

6.13 Exercises

1. Perform a von Neumann stability analysis on Eq. 6.3 and show that the stability condition is $\Delta t\,(\Delta x^{-2} + \Delta y^{-2}) \leq \tfrac{1}{2}$.

2. The equation of flow for a slightly compressible fluid in a 2-D homogeneous isotropic reservoir assuming gravity terms are zero and μ and c are constants is

$$p_{xx} + p_{yy} = \frac{1}{\alpha} p_t; \quad \alpha = \frac{k}{\phi \mu c}.$$

Suppose R is defined by $0 \leq x \leq 1$, $0 \leq y \leq 1$ and the following conditions hold:

(1) $p = 1$ when $t = 0$
(2) $u_y = 0$ when $y = 0$
(3) $u_x = 0$ when $x = 0$
(4) $u = 0$ when $x = 1$
(5) $u = 0$ when $y = 1$

(a) Assume $N_x = N_y = 14$, $\alpha = 1$. Develop a numerical solution for $p(x,y,t)$ using Peaceman-Rachford ADI (noniterative). Simulate this system using the following time steps: $\Delta t = 0.001$ (6 times); 0.002 (4 times); 0.003 (2 times); 0.005 (4 times); 0.01 (2 times); 0.02 (4 times); 0.03 (2 times); 0.05 (4 times); 0.1 (6 times); and 0.25 (2 times) for a total time $t = 1.5$. Print at least four decimal digits.

(b) The analytical solution to this problem is given by

$$p(x,y,t) = \sum_{m,n=1}^{\infty} \frac{16}{mn\pi^2} \sin\left(\frac{m\pi x}{2}\right) \sin\left(\frac{n\pi x}{2}\right) e^{-(n^2+m^2)\pi/4 t}.$$

Compare your numerical solution and that determined from the expression above for $x = y = 0.5$, $t = 0.1$.

(c) Use the results you obtained at $t = 0.06$ as the initial condition and simulate to $t = 0.1$ in two steps with $\Delta t = 0.02$. Repeat this calculation using ten steps with $\Delta t = 0.004$. Determine the maximum of the absolute differences between both calculations. What is the reason for these differences? How does the last calculation compare with the analytical solution for $x = 0.5$, $y = 0.5$, $t = 0.1$?

3. Establish Eq. 6.42.

4. Show that the overall equation for the Douglas-Rachford method for solving Eq. 6.41 is

$$(\Delta_x^2 + \Delta_y^2 + \Delta_z^2) u^{n+1} = \frac{u^{n+1} - u^n}{\Delta t} + (\Delta_x^2 \Delta_y^2 + \Delta_x^2 \Delta_z^2 + \Delta_y^2 \Delta_z^2)(u^{n+1} - u^n)\Delta t$$
$$- \Delta t^2 [\Delta_x^2 \Delta_y^2 \Delta_z^2 (u^{n+1} - u^n)].$$

5. Derive Thomas' tridiagonal algorithm from the band algorithm, Eqs. 6.71-74.

6. Determine the **L** and **U** factors for the matrix below.

$$A = \begin{bmatrix} 2 & 8 & 4 & 2 \\ 9 & 3 & 4 & 6 \\ 2 & 6 & -5 & 2 \\ 6 & 2 & 1 & 1 \end{bmatrix}.$$

(refer to Appendix A.6.10)

7. Using graph theory, determine the fill for a matrix problem $Ax = b$ where A is 5×5 having nonzeros only on the diagonal and on the last row and last column. Assume the unknowns are eliminated in the following orders:

   (1) x_1, x_2, x_3, x_4
   (2) x_5, x_1, x_2, x_3
   (3) x_1, x_5, x_2, x_3
   (4) x_2, x_4, x_3, x_1

8. Suppose a single-phase 2-D flow problem is to be simulated for the reservoir below with grid block indices as shown.

				1	2
3	4	5	6		
7	8				

   (a) What fill occurs for normal grid ordering of the reservoir blocks?
   (b) For alternate grid ordering? (Blocks 7 and 8 become 3 and 4, etc.; i.e., alternate rows of grid blocks are skipped and numbered on a second pass.)
   (c) Diagonal ordering? (All blocks sequentially numbered diagonally.)
   (d) Alternate diagonal ordering?

9. Refer to Appendices A.6.7 and A.6.8 and determine which of the matrices below are reducible.

   (a) $\begin{bmatrix} -1 & 0 & 3 & 1 \\ 3 & 2 & 1 & -2 \\ 2 & 0 & 0 & 4 \\ 0 & 0 & 1 & -1 \end{bmatrix}$

   (b) $\begin{bmatrix} -1 & 0 & 3 & 1 \\ 3 & 2 & 1 & -2 \\ 0 & 2 & 0 & 4 \\ 0 & 0 & 1 & -1 \end{bmatrix}$

   (c) $\begin{bmatrix} 0 & 1 & 1 \\ 1 & 0 & -1 \\ 2 & 1 & 0 \end{bmatrix}$

   (d) $\begin{bmatrix} 1 & 2 & 0 \\ 1 & 1 & 1 \\ 2 & 1 & 0 \end{bmatrix}$

(e) $$\begin{bmatrix} 2 & -1 & & & & \\ -1 & 2 & -1 & & & \\ & \cdot & \cdot & \cdot & & \\ & & & 2 & -1 \\ & & & -1 & 2 \end{bmatrix}$$

(f) $$\begin{bmatrix} 2 & -1 & & & & \\ 0 & 2 & & \cdot & & \\ & \cdot & \cdot & & & \\ & & & \cdot & 2 & -1 \\ & & & & 0 & 2 \end{bmatrix}$$

10. An incompressible flow regime in 2 − D is given by
$$u_{xx} + u_{yy} = -2, \ (x,y)\epsilon[0,1] \times [0,1]$$
$$u(x,y) = 0 \text{ on the boundary.}$$
   (a) Express this in the form $\mathbf{Au} = \mathbf{b}$, i.e., display $\mathbf{A}$, $\mathbf{u}$ and $\mathbf{b}$ when $\Delta x = \Delta y = \frac{1}{3}$.
   (b) Show that $\mathbf{A}$ is irreducible and diagonally dominant (see Appendix A.6.9).
   (c) Derive the point Jacobi and SOR iterative matrices.

11. (a) Find the Gauss-Seidel iteration matrix for the problem
$$5x_1 - 2x_2 = 5$$
$$-x_1 + 3x_2 = 3.$$
   (b) Iterate for 7 iterations and tabulate your results using for a first guess, $x_1 = x_2 = 1.0$. Also record the residuals.
   (c) Note that by solving the first equation for x_1 and the second for x_2 we get
$$x_1 = \frac{5 + 2x_2}{5} = 1 + \frac{2}{5} x_2$$
$$x_2 = \frac{3 + x_1}{3} = 1 + \frac{1}{3} x_1$$
   from which we get the iterative scheme
$$x_1^{(k+1)} = 1 + \tfrac{2}{5} x_2^{(k)}$$
$$x_2^{(k+1)} = 1 + \tfrac{1}{3} x_1^{(k+1)}.$$
   Show that this is identical to the Gauss-Seidel procedure.
   (d) Repeat the problem in part (a) using SOR with $\omega = 1.1$. Instead of developing the iteration matrix, an equivalent procedure is
$$x_1^{(k+1)} = [1 + \tfrac{2}{5} x_2^{(k)} - x_1^{(k)}]\omega + x_1^{(k)}$$
$$x_2^{(k+1)} = [1 + \tfrac{1}{3} x_1^{(k+1)} - x_2^{(k)}]\omega + x_2^{(k)}.$$
   (e) Is the value of ω used in part (d) above or below the optimum?
   (f) Determine the exact solution for the problem in part (a).

6.14 References

1. Settari, A. and Aziz, K.: "Use of Irregular Grid in Reservoir Simulation," *Soc. Pet. Eng. J.* (April 1972) 103.

2. Peaceman, D.W. and Rachford, H.H.: "The Numerical Solution of Parabolic and Elliptic Differential Equations," *J. Soc. Ind. Appl. Math.* (March 1955) **3**, No. 1, 28–41.
3. Douglas, J., Jr. and Dupont, T.: "Alternating Direction Galerkin Methods on Rectangles," *Numerical Solution of Partial Differential Equations-II*, Academic Press, New York City (1971).
4. Farrar, R.L.: "Gas Reservoir Simulation by Alternating Direction Galerkin Methods," PhD dissertation, Univ. of Tulsa, Tulsa, OK (1975).
5. Kebaili, A. and Thomas, G.W.: "A Two-Phase Coning Model Using Alternating Direction Galerkin Procedure," paper SPE 5724 presented at the 4th Symposium on Numerical Simulation of Reservoir Performance, Los Angeles, Feb. 19–20, 1976.
6. Thomas, G.W.: "Development and Application of Variational Methods to Reservoir Simulation," presented at the Congreso Panamericano de Ingenieria del Petroleo, Mexico City, March 19–23, 1979.
7. Ramey, H.J., Jr.: "Commentary on the Terms 'Transmissibility' and 'Storage'," *J. Pet. Tech.* (March 1975) 294.
8. Wood, W.L.: "Control of Crank-Nicolson Noise in the Numerical Solution of the Heat Conduction Equation," *Int'l. J. for Num. Meth. in Eng.* (July 1977) **11**, No. 17, 1059.
9. Bangia, V.K., Bennett, C., Reynolds, A., Raghavan, and Thomas, G.W.: "Alternating Direction Collocation Methods for Simulating Reservoir Performance," paper SPE 7414 presented at the SPE 53rd Annual Technical Conference and Exhibition, Houston, Oct. 1–4, 1978.
10. Peaceman, D.W.: *Fundamentals of Numerical Reservoir Simulation*, Elsevier North-Holland, Inc., New York City (1977).
11. Douglas, J.: "A Survey of Numerical Methods for Parabolic Differential Equations," *Advances in Computers*, second edition, Academic Press, New York City (1961) 2.
12. Douglas, J. and Rachford, H.H.: "On the Numerical Solution of Heat Conduction Problems in Two or Three Space Variables," *Trans. Am. Math. Soc.* (May–Aug. 1956) **82**, 421.
13. Brian, P.L.T.: "A Finite Difference Method of High-Order Accuracy for the Solution of Three-Dimensional Transient Heat Conduction Problems," *AICHE J.* (Sept. 1961) **7**, No. 3, 367–371.
14. Douglas, J.: "Alternating Direction Methods for Three Space Variables," *Numerische Matematikk* (1962) **4**, 41.
15. Woo, P. T., Roberts, S.J., and Gustavson, F.G.: "Application of Sparse Matrix Techniques in Reservoir Simulation," paper SPE 4544 presented at the SPE 48th Annual Meeting, Las Vegas, Sept. 30–Oct. 3, 1973.
16. Gustavson, F.G., Liniger, W.M., and Willoughby, R.A.: "Symbolic Generation of an Optimal Crout Algorithm for Sparse Systems of Linear Equations," *J. ACM* (Jan. 1970) **17**, 1,87.
17. Ogbuobiri, E.C., Tinney, W.F., and Walker, J.W.: "Sparsity-Directed Decomposition for Gaussian Elimination on Matrices," *IEEE Trans.* (Jan. 1970) **PAS-89**, No. 1, 141.
18. Price, H.S. and Coats, K.H.: "Direct Methods in Reservoir Simulation," *Trans.*, AIME (1974) **257**, 295.
19. George, A. and Liu, J.W-H.: *Computer Solution of Large Sparse Positive Definite Systems*, Prentice-Hall, Inc., Englewood Cliffs (1981).
20. Tinney, W.F. and Walker, J.W.: "Direct Solutions of Sparse Network Equations by Optimally Ordered Triangular Factorization," *Proc. IEEE* (Nov. 1967) **55**, 1801.

21. Berry, R.D.: "An Optimum Ordering of Electronic Circuit Equations for a Sparse Matrix Solution," *IEEE Trans.* (Jan. 1971) **CT-18,** 40.
22. Stone, H.L.: "Iterative Solutions of Implicit Approximations of Multi-Dimensional Partial Differential Equations," *SIAM J. on Num. Anal.* (Sept. 1968) **5,** 530.
23. Dupont, T., Kendall, R.P., and Rachford, H.H.: "An Approximate Factorization Procedure for Solving Self-Adjoint Elliptic Difference Equations," *SIAM J. on Num. Anal.* (Sept. 1968) **5,** 539.
24. Weinstein, H.G., Stone, H.L., and Kwan, T.V.: "Iterative Procedure for Solution of Systems of Parabolic and Elliptic Equations in Three Dimensions," *Ind. and Eng. Chem. Fund* (May 1969) **8,** 282.
25. Weinstein, H.G., Stone, H.L., and Kwan, T.V.: "Simultaneous Solution of Multiphase Reservoir Flow Problems," *Soc. Pet. Eng. J.* (June 1970) 99.
26. Varga, R.S.: *Matrix Iterative Analysis,* Prentice-Hall, Inc., Englewood Cliffs (1962).
27. Burden, R.L., Faires, J.D., and Reynolds, A.C.: *Numerical Analysis,* Prindle, Weber & Schmidt, Boston (1978).
28. Rigler, A.K.: "Estimation of the Successive Over-Relaxation Factor," *Math Comp.* (April 1965) **19,** No. 90, 302–307.
29. Carre, B.A.: "The Determination of the Optimum Accelerating Factor for Successive Overrelaxation," *Comput. J.* (April 1961) **4,** 73.
30. Hageman, L.A. and Kellogg, R.B.: "Estimating Optimum Overrelaxation Parameters," *Math. Comp.* (Jan. 1968) **22,** 60.
31. Kebaili, A.: "Tensor Product Methods for Determining Optimum Relaxation Factors," MS Thesis, Univ. of Tulsa, Tulsa, OK (1973).

7
Multiphase Flow

These motions everywhere in nature
Must surely be
The circulations of God.

Thoreau

7.1 Introduction

The principles outlined in the preceding chapter can be readily extended to multiphase, multidimensional flow. We begin our discussion with two-phase incompressible flow and follow this with a treatment of three-phase compressible flow. We confine ourselves to heterogeneous, anisotropic media in 3-D. Actually, the three-phase, compressible model represents the most general case for such an environment in a black-oil reservoir. A simulator constructed to treat the latter can equally well handle the incompressible case by supplying appropriate input data. Furthermore, flow regimes in either 1, 2, or 3-D can be similarly treated.

We outline three methods of solution for the two-phase flow problem, a simultaneous solution technique, the "leap-frog" approach, and an implicit pressure, explicit saturation procedure (IMPES). For the three-phase compressible problem, we discuss the IMPES method, a fully implicit approach, and an adaptive implicit method.

7.2 Two-Phase Flow

For a two-phase flow system, we consider the water and oil equations given by Eqs. 4.44 and 4.45 and treat first a simultaneous solution procedure similar to that given by Coats et al.[1]

$$\nabla \cdot \left\{ \frac{k\, k_{rw}}{\mu_w B_w} \right\} \nabla \Phi_w + Q_w = \phi \frac{\partial}{\partial t}\left(\frac{S_w}{B_w}\right) \tag{7.1}$$

$$\nabla \cdot \left\{ \frac{k\, k_{ro}}{\mu_o B_o} \nabla \Phi_o \right\} + Q_o = \phi \frac{\partial}{\partial t}\left(\frac{S_o}{B_o}\right) \tag{7.2}$$

For simplicity, we associate only plus signs with the source/sink terms since the choice is arbitrary. If we assume the fluids are incompressible then B_o and B_w are constants, and

$$\nabla \cdot \left\{ \frac{k\, k_{rl}}{\mu_l} \nabla \Phi_l \right\} + B_l\, Q_l = \phi \frac{\partial S_l}{\partial t},\ l = o,\ w. \tag{7.3}$$

We also invoke the following relations:

$$S_o + S_w = 1 \tag{7.4}$$

$$P_{cwo} = p_o - p_w \tag{7.5}$$

$$\Phi_o = p_o - \gamma_o d \tag{7.6}$$

$$\Phi_w = p_w - \gamma_w d \tag{7.7}$$

Then,

$$P_{cwo} = \Phi_o - \Phi_w - \delta\gamma d;\quad \delta\gamma \equiv \gamma_w - \gamma_o. \tag{7.8}$$

At the water-oil contact (WOC) $P_{cwo} = 0$; thus we have the initial conditions

$$\Phi_o = \Phi_w + (\delta\gamma) d_{woc}. \tag{7.9}$$

7.2.1 Simultaneous Solution Method

Let $S_w = S$ such that $S_o = 1 - S$ from Eq. 7.4. The right-hand sides of Eq. 7.3 become $\phi\, \partial S/\partial t$ and $-\phi\, \partial S/\partial t$, respectively. Now

$$\begin{aligned}\frac{\partial S}{\partial t} &= \frac{\partial S}{\partial P_{cwo}}\frac{\partial P_{cwo}}{\partial t} = S'\frac{\partial P_{cwo}}{\partial t} \\ &= S'\left\{\frac{\partial \Phi_o}{\partial t} - \frac{\partial \Phi_w}{\partial t}\right\}\end{aligned} \tag{7.10}$$

using Eq. 7.8. If Eq. 7.10 is substituted in Eq. 7.3 then,

Multiphase Flow

$$\nabla \cdot \left\{ \frac{k\, k_{rw}}{\mu_w} \nabla \Phi_w \right\} + B_w Q_w = -S'\phi \frac{\partial \Phi_w}{\partial t} + S'\phi \frac{\partial \Phi_o}{\partial t} \qquad (7.11)$$

$$\nabla \cdot \left\{ \frac{k\, k_{ro}}{\mu_o} \nabla \Phi_o \right\} + B_o Q_o = S'\phi \frac{\partial \Phi_w}{\partial t} - S'\phi \frac{\partial \Phi_o}{\partial t} \qquad (7.12)$$

which have the finite difference approximations

$$\Delta T_w \Delta \Phi_w + B_w q_w = G \Delta_t \Phi_w - G \Delta_t \Phi_o \qquad (7.13)$$

$$\Delta T_o \Delta \Phi_o + B_o q_o = -G \Delta_t \Phi_w + G \Delta_t \Phi_o \qquad (7.14)$$

where

$$G \equiv -\frac{S'\phi \Delta x_i\, \Delta y_j\, \Delta z_k}{\Delta t\, 5.61416}$$

$$q_w \equiv \frac{\Delta x_i\, \Delta y_j\, \Delta z_k\, Q_w}{5.61416}$$

$$q_o \equiv \frac{\Delta x_i\, \Delta y_j\, \Delta z_k\, Q_o}{5.61416}$$

$$\Delta_t \Phi_l \equiv \Phi_l^{n+1} - \Phi_l^n$$

$$\Delta T_l \Delta \Phi_l \equiv \Delta_x T_{lx} \Delta_x \Phi_l + \Delta_y T_{ly} \Delta_y \Phi_l + \Delta_z T_{lz} \Delta_z \Phi_l, \quad l = o, w$$

$$(T_{ox})_{i \pm 1/2} \equiv \left\{ \frac{\alpha k\, k_{ro}}{\Delta x_i \mu_o} \right\}_{i \pm 1/2}^n \Delta y_j\, \Delta z_k$$

and $\alpha \equiv 0.001127$, for k in md., Δx_i, Δy_j, Δz_k in ft. and μ_o in cp. T_{ly} and T_{lz}, $l = o, w$, have definitions similar to T_{ox}.

We assume the transmissibilities are evaluated at time level n; i.e., they are computed explicitly. The transmissibilities contain saturation dependent terms (k_{rl}). These are evaluated at the half-nodal points by a weighting technique, i.e.,

$$(k_{rl})_{i+1/2} = w(k_{rl})_{\text{upstream}} + (1-w)(k_{rl})_{\text{downstream}} \qquad (7.15)$$

where $0 \leq w \leq 1$. If $w = \frac{1}{2}$ then Eq. 7.15 produces the arithmetic average. In Fig. 7.1, the ith block is the upstream block while the $(i+1)^{st}$ block is downstream.

To obtain physically realistic results it is necessary to employ upstream

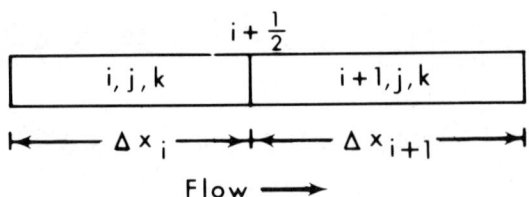

Fig. 7.1 Upstream and Downstream Blocks.

weighting on k_{rl}, i.e., $w = 1.0$ for the situation in Fig. 7.1. Moreover, for 1-D problems in the absence of capillary pressure, upstream weighting insures stability for Δt sufficiently small.[2] Thus, if $w = 1$, *single-point* upstream weighting is applied to the relative permeability which is first order correct. A *two-point* weighting scheme which is second order correct has been presented by Todd. et al.[3] The absolute permeability k is computed as a harmonic average,

$$\left(\frac{k}{\Delta x}\right)_{i+1/2} = \frac{2\, k_i k_{i+1}}{k_i\, \Delta x_{i+1} + k_{i+1}\, \Delta x_i}. \quad (7.16)$$

In incompressible flow, the viscosities μ_o and μ_w are constants. In the compressible flow case they are pressure dependent functions. They can be satisfactorily computed at the block interfaces by averaging between blocks.

If we expand Eqs. 7.13 and 7.14 we obtain

$$\Delta T_w \Delta \Phi_w^{n+1} - G\Phi_w^{n+1} + G\Phi_o^{n+1} = -B_w q_w^n - G\Phi_w^n + G\Phi_o^n \quad (7.17)$$

$$\Delta T_o \Delta \Phi_o^{n+1} + G\Phi_w^{n+1} - G\Phi_o^{n+1} = -B_o q_o^n + G\Phi_w^n - G\Phi_o^n \quad (7.18)$$

We formulate the solution of Eqs. 7.17–7.18 in terms of the Douglas-Rachford iterative procedure presented in chapter 6. For the water equation (Eq. 7.17) we have

x-sweep:

$$\Delta_x T_{wx} \Delta_x \Phi_w^{k+1/3} - G\Phi_w^{k+1/3} + G\Phi_o^{k+1/3} + \Delta_y T_{wy} \Delta_y \Phi_w^k + \Delta_z T_{wz} \Delta_z \Phi_w^k$$
$$= -B_w q_w^n - G\Phi_w^n + G\Phi_o^n + a \quad (7.19)$$

y-sweep:

$$\Delta_x T_{wx} \Delta_x \Phi_w^{k+1/3} - G\Phi_w^{k+2/3} + G\Phi_o^{k+2/3} + \Delta_y T_{wy} \Delta_y \Phi_w^{k+2/3} + \Delta_z T_{wz} \Delta_z \Phi_w^k$$
$$= -B_w q_w^n - G\Phi_w^n + G\Phi_o^n + b \quad (7.20)$$

z-sweep:

$$\Delta_x T_{wx} \Delta_x \Phi_w^{k+1/3} - G\Phi_w^{k+1} + G\Phi_o^{k+1} + \Delta_y T_{wy} \Delta_y \Phi_w^{k+2/3} + \Delta_z T_{wz} \Delta_z \Phi_w^{k+1}$$
$$= -B_w q_w^n - G\Phi_w^n + G\Phi_o^n + c \quad (7.21)$$

where

$$a = \sigma_k(\Sigma T_w)(\Phi_w^{k+1/3} - \Phi_w^k)$$
$$b = \sigma_k(\Sigma T_w)(\Phi_w^{k+2/3} - \Phi_w^k)$$
$$c = \sigma_k(\Sigma T_w)(\Phi_w^{k+1} - \Phi_w^k)$$

As in chapter 6, ΣT_w means the sums of the water transmissibilities over the faces of a grid block. Similar expressions to Eq. 7.21 are written for Eq. 7.18. We employ the following definitions to minimize round-off and cancellation errors:

$$PX = \Phi_w^{k+1/3} - \Phi_w^k \quad (7.22a)$$

$$PY = \Phi_w^{k+2/3} - \Phi_w^k \quad (7.22b)$$

$$PZ = \Phi_w^{k+1} - \Phi_w^k \quad (7.22c)$$

$$RX = \Phi_o^{k+1/3} - \Phi_o^k \quad (7.22d)$$

$$RY = \Phi_o^{k+2/3} - \Phi_o^k \quad (7.22e)$$

$$RZ = \Phi_o^{k+1} - \Phi_o^k \quad (7.22f)$$

If we use Eqs. 7.22a–7.22c in Eqs. 7.19–7.21 then,

x-sweep:

$$\Delta_x T_{wx} \Delta_x PX + \Delta T_w \Delta \Phi_w^k - G\,PX + G\,RX$$
$$= -B_w q_w^n + G(\Phi_w^k - \Phi_w^n) - G(\Phi_o^k - \Phi_o^n) + \sigma_k(\Sigma T_w)\,PX \quad (7.23)$$

y-sweep:

$$\Delta_x T_{wx} \Delta_x PX + \Delta T_w \Delta\Phi_w^k - G\,PY + G\,RY + \Delta_y T_{wy} \Delta_y PY$$
$$= -B_w q_w^n + G(\Phi_w^k - \Phi_w^n) - G(\Phi_o^k - \Phi_o^n) + \sigma_k(\Sigma T_w)\,PY \quad (7.24)$$

z-sweep:

$$\Delta_x T_{wx} \Delta_x PX + \Delta_y T_{wy} \Delta_y PY + \Delta_z T_{wz} \Delta_z PZ + \Delta T_w \Delta\Phi_w^k - G\,PZ + G\,RZ$$
$$= -B_w q_w^n + G(\Phi_w^k - \Phi_w^n) - G(\Phi_o^k - \Phi_o^n) + \sigma_k(\Sigma T_w)\,PZ \quad (7.25)$$

Again, Eqs. 7.22d–7.22f result in a similar expression for the oil equation. Finally, we employ the following definitions:

$$RWX \equiv \Delta T_w \Delta \Phi_w^k + B_w q_w^n - G(\Phi_w^k - \Phi_w^n) + G(\Phi_o^k - \Phi_o^n)$$

$$ROX \equiv \Delta T_o \Delta \Phi_o^k + B_o q_o^n + G(\Phi_w^k - \Phi_w^n) - G(\Phi_o^k - \Phi_o^n)$$

$$RWY \equiv (G + \sigma_k \Sigma T_w)\, PX - GRX$$

$$ROY \equiv (G + \sigma_k \Sigma T_o)\, RX - GPX$$

$$RWZ \equiv (G + \sigma_k \Sigma T_w)\, PY - GRY$$

$$ROZ \equiv (G + \sigma_k \Sigma T_o)\, RY - GPY$$

and ultimately obtain

x-sweep:

$$\Delta_x T_{wx} \Delta_x PX - (G + \sigma_k \Sigma T_w)\, PX + GRX = -RWX \qquad (7.26a)$$

$$\Delta_x T_{ox} \Delta_x RX - (G + \sigma_k \Sigma T_o)\, RX + GPX = -ROX \qquad (7.26b)$$

y-sweep:

$$\Delta_y T_{wy} \Delta_y PY - (G + \sigma_k \Sigma T_w)\, PY + GRY = -RWY \qquad (7.27a)$$

$$\Delta_y T_{oy} \Delta_y RY - (G + \sigma_k \Sigma T_o)\, RY + GPY = -ROY \qquad (7.27b)$$

z-sweep:

$$\Delta_z T_{wz} \Delta_z PZ - (G + \sigma_k \Sigma T_w)\, PZ + GRZ = -RWZ \qquad (7.28a)$$

$$\Delta_z T_{oz} \Delta_z RZ - (G + \sigma_k \Sigma T_o)\, RZ + GPZ = -ROZ \qquad (7.28b)$$

for both the water and oil equations. The right-hand sides of Eqs. 7.26–7.28 are residuals. Compare the definitions of RWX and ROX to Eqs. 7.17 and 7.18, and note that these should approach zero when convergence is achieved. Thus, satisfactory normalized closure criteria are

$$\left| \frac{\Sigma B_w q_w - \Sigma RWX}{\Sigma B_w q_w} \right| \leq \epsilon_w \qquad (7.29)$$

Multiphase Flow

$$\left| \frac{\Sigma B_o q_o - \Sigma ROX}{\Sigma B_o q_o} \right| \leq \epsilon_o \qquad (7.30)$$

where the sums are over all active grid blocks.

Eqs. 7.26–7.28 constitute a bi-tridiagonal system that can be solved by an algorithm given by Douglas, Peaceman, and Rachford.[4] In applying the procedure, the iteration parameters are varied cyclically until convergence is achieved. To compute the values of G, we must have S' which is updated after every iteration using

$$(S')^{k+1} = \frac{k}{N_i} \left(\frac{S^k - S^n}{P^k_{cwo} - P^n_{cwo}} \right) + (1 - k/N_i)(S')^k \qquad (7.31)$$

where k is the iteration level and N_i is the total number of iterations per cycle.

7.2.2 Leap-Frog Technique

Douglas, Peaceman, and Rachford[4] propose an alternative procedure to that given in Eqs. 7.26–7.28 known as the "leap-frog" technique. Its principal advantage is that substantially less work is required to arrive at a solution. The idea is to eliminate one of the two unknowns in Eqs. 7.11 and 7.12 and thus reduce the computing time. Define two new independent variables: $P \equiv (\Phi_o + \Phi_w)/2$ and $R \equiv (\Phi_o - \Phi_w)/2$ and the functions,

$$M \equiv k \left\{ \frac{k_{ro}}{\mu_o} + \frac{k_{rw}}{\mu_w} \right\}, \qquad N \equiv k \left\{ \frac{k_{ro}}{\mu_o} - \frac{k_{rw}}{\mu_w} \right\}.$$

Now add and subtract Eqs. 7.11 and 7.12 using these definitions to get

$$\nabla \cdot M \nabla P + \nabla \cdot N \nabla R + B_w Q_w + B_o Q_o = 0 \qquad (7.32)$$

$$\nabla \cdot M \nabla R + \nabla \cdot N \nabla P + B_o Q_o - B_w Q_w = -4 \phi S' \frac{\partial R}{\partial t} \qquad (7.33)$$

The finite difference approximations of these equations are

$$\Delta M^m \Delta P^{m+1} + \Delta N^m \Delta R^m + B_w q_w^m + B_o q_o^m = 0 \qquad (7.34)$$

$$\Delta M^m \Delta R^{m+1} + \Delta N^m \Delta P^{m+1} + B_o q_o^m - B_o q_o^m = -\frac{4 \phi S'}{\Delta t} \{R_{ijk}^{m+1} - R_{ijk}^m\} \qquad (7.35)$$

Eq. 7.34 is used to determine P at the $(m + 1)^{st}$ time level using known values of R, N, M from the previous level. Then p^{m+1} is used in Eq. 7.35

to solve for the remaining unknown R^{m+1}. Douglas et al[4] experienced an approximate four-fold decrease in computing time using this technique.

7.2.3 IMPES Formulation

The implicit pressure-explicit saturation (IMPES) technique and its variations are widely used in commercial reservoir simulators. This method involves eliminating the saturation terms from the flow equations to get an equation that involves only one dependent variable, usually potential or pressure. This is solved implicitly. The saturation is then computed explicitly by referring back to one of the flow equations. The technique was originally proposed by Sheldon et al.[5] and has been extended by Stone and Garder.[6]

Except in reservoir situations having a high degree of stratification, capillary forces may not have a significant effect on the computed results. The methods discussed thus far always require a non-constant value of capillary pressure because $\Delta_t S$ terms are replaced by $S' \Delta_t P_c$, and for $P_c =$ constant, $S' = \infty$ which, of course, is invalid. Thus, in those cases where the capillary forces have little effect on the results, we cannot set $P_c = 0$ in the models treated in the preceding sections. This means there will always be the additional overhead in computer time required to evaluate S'. One advantage of the IMPES method is that zero capillary pressures can be employed and such simulators generally run faster. We briefly sketch the technique below. More detail is provided when we treat three-phase flow.

Adding Eqs. 7.13 and 7.14 yields

$$\Delta T_w \Delta \Phi_w + \Delta T_o \Delta \Phi_o = -B_w q_w - B_o q_o \equiv -\tilde{q}. \tag{7.36}$$

We solve Eq. 7.8 for Φ_o and write

$$\Delta \Phi_o = \Delta \Phi_w + \Delta (P_{cwo} + \delta \gamma d) \equiv \Delta \Phi + \Delta R \tag{7.37}$$

which when substituted in Eq. 7.36 gives

$$\Delta \tilde{T} \Delta \Phi + \Delta T_o \Delta R = -\tilde{q} \tag{7.38}$$

where $\tilde{T} = T_o + T_w$. The transmissibilities, T_o and T_w, R and $\tilde{q}$ are evaluated explicitly, i.e., at time level n. Thus we have the elliptic form

$$\Delta \tilde{T}^n \Delta \Phi^{n+1} = -\Delta T_o^n \Delta R^n - \tilde{q}^n \equiv G \tag{7.39}$$

in the single unknown Φ. The Douglas-Rachford ADI formulation yields

$$x: \Delta_x \tilde{T}_x \Delta_x PX - \sigma_k (\Sigma \tilde{T}) PX = RX \tag{7.40}$$

$$y: \Delta_y \tilde{T} \Delta_y PY - \sigma_k(\Sigma \ \tilde{T}) PY = RY \qquad (7.41)$$

$$z: \Delta_z \tilde{T} \Delta_z PZ - \sigma_k(\Sigma \ \tilde{T}) PZ = RZ \qquad (7.42)$$

where PX, PY, PZ are defined in Eqs. 7.22a–7.22c and $RX = G - \Delta \tilde{T} \Delta \Phi^k$, $RY = -\sigma_k(\Sigma \ \tilde{T}) PX$, $RZ = -\sigma_k(\Sigma \ \tilde{T}) PY$. The steps involved in arriving at Eqs. 7.40–7.42 are identical to those used to get Eqs. 7.26–7.28. At the end of each iteration, we update the potential using $\Phi^{k+1} = \Phi^k + PZ$. Convergence is achieved when

$$\left| \frac{\Sigma \ RX}{\Sigma B_w q_w} \right| \leq \epsilon$$

since RX is identical to Eq. 7.39 when $\Phi^k \to \Phi^{n+1}$. Since Φ is the water potential, we also get the oil potential at the new time level using the capillary-pressure relationship in Eq. 7.8. If $P_{cwo} = 0$, then they differ only by the gravity term. The water saturation is computed using the explicit formula

$$S_w^{n+1} = S_w^n + \frac{\Delta t}{V_p} \{\Delta T_w \, \Delta \Phi_w^{n+1} + B_w q_w\} \qquad (7.43)$$

and the oil saturation is obtained by difference, i.e., $S_o^{n+1} = 1 - S_w^{n+1}$. Conversely, one can use an explicit formula for S_o^{n+1} and compute S_w^{n+1} by difference. As an alternative to an ADI approach, Eq. 7.39 could be solved by any of the solution techniques discussed in chapter 6. The matrix generated by the left-hand side is banded with each entry a single element, identical in structure to that coming from a single-phase flow problem.

7.3 Three-Phase Flow

The equations for three-phase flow in a black-oil reservoir are given in Eqs. 4.44–4.46. Their finite difference analogs are

$$\Delta \ T_w(\Delta p_w - \gamma_w \Delta d) + q_w = \frac{V_b}{\Delta t} \Delta_t(b_w S_w \phi) \qquad (7.44)$$

$$\Delta \ T_o(\Delta p_o - \gamma_o \Delta d) + q_o = \frac{V_b}{\Delta t} \Delta_t(b_o S_o \phi) \qquad (7.45)$$

$$\Delta R_s T_o(\Delta p_o - \gamma_o \Delta d) + \Delta T_g(\Delta p_g - \gamma_g \Delta d) + q_g + R_s q_o$$
$$= \frac{V_b}{\Delta t} \Delta_t(\phi \, b_g S_g + \phi \, R_s b_o S_o) \quad (7.46)$$

where

$$b_l = 1/B_l$$
$$q_l = \frac{\Delta x_i \Delta y_j \Delta z_k \, Q_l}{5.61416} \quad l = o, w, g$$
$$V_b = \frac{\Delta x_i \Delta y_j \Delta z_k}{5.61416}, \quad \text{the bulk volume}$$

The auxiliary relations are

$$S_o + S_w + S_g = 1 \quad (7.47)$$

$$P_{cwo} = p_o - p_w \quad (7.48)$$

$$P_{cgo} = p_g - p_o \quad (7.49)$$

7.3.1 IMPES Formulation

The development in this section follows that of Coats.[7] If we assume ϕ and B_l are pressure dependent then

$$\phi = \phi_0(1 + c_r p_w) \quad (7.50)$$

where c_r is the rock compressibility and ϕ_0 is the porosity at a datum, e.g., surface conditions (see chapter 3). The right-hand sides of Eqs. 7.44–7.46 then become

$$\frac{V_p}{\Delta t} \Delta_t [(1 + c_r \, p_w) \, b_l \, S_l], \, l = o, w \quad (7.51)$$

$$\frac{V_p}{\Delta t} \Delta_t [(1 + c_r \, p_w)(b_g \, S_g + R_s \, B_o \, S_o)] \text{ (for gas)} \quad (7.52)$$

V_p represents the pore volume of a block at the datum condition. As Coats[7] demonstrates a consistent expansion of Eqs. 7.51–7.52 requires that

$$\Delta_t(a \, b) = a^{n+1} \Delta_t b + b^n \Delta_t a. \quad (7.53)$$

Applying this definition to the right-hand side expressions we have (omitting $V_p/\Delta t$ for the time being)

$$\Delta_t \left[(1 + c_r\, p_w)\, b_l\, S_l \right] = (1 + c_r\, p_w)^{n+1} \Delta_t(b_l\, S_l) + (b_l\, S_l)^n \Delta_t(1 + c_r\, p_w)$$

which on further expansion becomes

$$(1 + c_r\, p_w)^{n+1} \{ b_l^{n+1} \Delta_t S_l + S_l^n \Delta_t b_l \} + (b_l\, S_l)^n \Delta_t(c_r p_w).$$

Now since $b_l = b_l(p)$ then $\Delta_t b_l = \Delta_p b_l \Delta_t p_l$ where

$$\Delta_p b_l = (b_l^{n+1} - b_l^n)/(p_l^{n+1} - p_l^n) \equiv b_l'.$$

Finally, for the right-hand sides of Eqs. 7.44 and 7.45,

$$\Delta_t \left[(1 + c_r p_w)\, b_l S_l \right] = (1 + c_r p_w)^{n+1} \{ b_l^{n+1} \Delta_t S_l + S_l b_l' \Delta_t p_l \}$$
$$+ (b_l S_l)^n \Delta_t (c_r p_w), \quad l = o, w. \quad (7.54)$$

For the right-hand side of Eq. 7.46 we have

$$\Delta_t \left[(1 + c_r p_w)\, b_g S_g + (1 + c_r p_w)\, R_s b_o S_o \right] = (1 + c_r p_w)^{n+1} \{ b_g^{n+1} \Delta_t S_g$$
$$+ S_g b_g' \Delta_t p_g \} + (b_g S_g)^n \Delta_t (c_r p_w) + \Delta_t \left[(1 + c_r p_w)\, R_s b_o S_o \right]$$
$$= E_1 + (1 + c_r p_w)^{n+1} \Delta_t(R_s b_o S_o) + (R_s b_o S_o)^n \Delta_t(c_r p_w)$$

where

$$E_1 \equiv (1 + c_r p_w)^{n+1} \{ b_g^{n+1} \Delta_t S_g + S_g b_g' \Delta_t p_g \} + (b_g S_g)^n \Delta_t(c_r p_w).$$

Further expansion of the terms having triple factors yields finally

$$\Delta_t \left[(1 + c_r p_w) b_g S_g + (1 + c_r p_w)\, R_s b_o S_o \right]$$
$$= (1 + c_r p_w)^{n+1} \{ b_g^{n+1} \Delta_t S_g + S_g b_g' \Delta_t p_g + R_s^{n+1} b_o^{n+1} \Delta_t(S_o)$$
$$+ b_o' R_s^{n+1} S_o^n \Delta_t p_o + (b_o S_o)^n R_s' \Delta_t p_o \} + (b_g S_g + R_s b_o S_o)^n \Delta_t\, c_r p_w \quad (7.55)$$

where we have used the definitions

$$\Delta_t R_s = \Delta_p R_s \Delta_t p_o \equiv R_s' \Delta_t p_o$$

$$\Delta_t b_g = \Delta_p b_g \Delta_t p_g \equiv b_g' \Delta_t p_g.$$

The final form for the right-hand sides is then

$$\frac{V_p}{\Delta t} \Delta_t[(1 + c_r p_w) b_w S_w] = \alpha_{10} \Delta_t p + \sum_{m=1}^{3} \alpha_{1m} \Delta_t S_m \quad (7.56)$$

$$\frac{V_p}{\Delta t}\Delta_t[(1+c_r p_w)b_o S_o] = \alpha_{20}\Delta_t p + \sum_{m=1}^{3}\alpha_{2m}\Delta_t S_m \tag{7.57}$$

$$\frac{V_p}{\Delta t}\Delta_t[(1+c_r p_w)(b_g S_g + R_s b_o S_o)] = \alpha_{30}\Delta_t p + \sum_{m=1}^{3}\alpha_{3m}\Delta_t S_m \tag{7.58}$$

where $\Delta_t p_g = \Delta_t p_w = \Delta_t p_o \equiv \Delta_t p$ i.e., we've neglected the change in P_c in a time step. In Eqs. 7.56–7.58 the index m takes on values 1, 2, 3 which corresponds to water *(w)*, oil *(o)*, and gas *(g)*, respectively. The values of alpha are defined as follows:

$$\alpha_{10} = \frac{V_p}{\Delta t}[(1+c_r p_w^{n+1})\,S_w^n\,b_w' + (b_w S_w)^n c_r]$$

$$\alpha_{11} = \frac{V_p}{\Delta t}(1+c_r p_w^{n+1})\,b_w^{n+1};\; \alpha_{12} = \alpha_{13} = 0$$

$$\alpha_{20} = \frac{V_p}{\Delta t}[(1+c p_w^{n+1})\,S_o^n b_o' + (b_o S_o)^n c_r]$$

$$\alpha_{22} = \frac{V_p}{\Delta t}(1+c_r p_w^{n+1})\,b_o^{n+1};\; \alpha_{21} = \alpha_{23} = 0$$

$$\alpha_{30} = \frac{V_p}{\Delta t}\{(1+c_r p_w^{n+1})[S_g^n b_g' + S_o^n (b_o R_s)'] + (b_g S_g + R_s b_o S_o)^n c_r\}$$

$$\alpha_{31} = 0$$

$$\alpha_{32} = \frac{V_p}{\Delta t}[(1+c_r p_w^{n+1})(R_s b_o)^{n+1}]$$

$$\alpha_{33} = \frac{V_p}{\Delta t}[(1+c_r p_w^{n+1})\,b_g^{n+1}]$$

Thus we have finally for Eqs. 7.44–7.46,

$$\Delta[T_w(\Delta p_w - \gamma_w \Delta d)] + q_w = \alpha_{10}\Delta_t p + \sum_{m=1}^{3}\alpha_{1m}\Delta_t S_m \tag{7.59}$$

$$\Delta[T_o(\Delta p_o - \gamma_o \Delta d)] + q_o = \alpha_{20}\Delta_t p + \sum_{m=1}^{3}\alpha_{2m}\Delta_t S_m \tag{7.60}$$

$$\Delta[T_g(\Delta p_g - \gamma_g \Delta d)] + \Delta[R_s T_o(\Delta p_o - \gamma_o \Delta d)] + q_g + R_s q_o$$

$$= \alpha_{30} \Delta_t p + \sum_{m=1}^{3} \alpha_{3m} \Delta_t S_m \qquad (7.61)$$

Multiply Eq. 7.59 by β_1, Eq. 7.60 by β_2, Eq. 7.61 by β_3 and add to get

$$\beta_1 \Delta[T_w(\Delta p_w - \gamma_w \Delta d)] + \beta_2 \Delta[T_o(\Delta p_o - \gamma_o \Delta d)] + \beta_3 \Delta[T_g(\Delta p_g - \gamma_g \Delta d)]$$
$$+ \beta_3 \Delta[R_s T_o(\Delta p_o - \gamma_o \Delta d)] + \beta_1 q_w + \beta_2 q_o + \beta_3(q_g + R_s q_o)$$
$$= \beta \Delta_t p + \beta_1 \Sigma \alpha_{1m} \Delta_t S_m + \beta_2 \Sigma \alpha_{2m} \Delta_t S_m + \beta_3 \Sigma \alpha_{3m} \Delta_t S_m \qquad (7.62)$$

where

$$\beta = \beta_1 \alpha_{10} + \beta_2 \alpha_{20} + \beta_3 \alpha_{30}.$$

If we expand the right-hand side of Eq. 7.62 and use $\Delta_t S_g = -\Delta_t S_o - \Delta_t S_w$ (which follows from Eq. 7.42) then we get

$$RHS = \beta \Delta_t p + (\beta_1 \alpha_{11} - \beta_3 \alpha_{33}) \Delta_t S_w + (\beta_2 \alpha_{22} + \beta_3 \alpha_{32} - \beta_3 \alpha_{33}) \Delta_t S_o. \qquad (7.63)$$

The coefficients of $\Delta_t S_w$ and $\Delta_t S_o$ are set to zero. This yields two expressions in the three unknowns β_i, $i = 1, 2, 3$. We select $\beta_1 = 1$ arbitrarily which implies that $\beta_2 = (\alpha_{11}/\alpha_{22})(1 - \alpha_{32}/\alpha_{33})$ and $\beta_3 = \alpha_{11}/\alpha_{33}$. From Eqs. 7.48 and 7.49 we have $\Delta p_o = \Delta p_w + \Delta P_{cwo} = \Delta p_g - \Delta P_{cgo} \equiv \Delta p$ which is used in the left-hand side of Eq. 7.62. Finally,

$$\Delta \tilde{T} \Delta p^{n+1} = \beta \Delta_t p + G \qquad (7.64)$$

where $\tilde{T} \equiv T_w + \beta_2 T_o + \beta_3(T_g + R_s T_o)$, $G \equiv \Delta(T_w \Delta P_{cwo}) - \beta_3 \Delta(T_g \Delta P_{cgo}) - \Delta[\gamma_w T_w + \beta_2 \gamma_o T_o + \beta_3(\gamma_g T_g + \gamma_o R_s T_o)] \Delta d - \tilde{q}$, $\tilde{q} \equiv q_w + \beta_2 q_o + \beta_3(q_g + R_s q_o)$ and all terms in G, $\tilde{q}$ and $\tilde{T}$, are evaluated at time level n. Eq. 7.64 involves a single unknown, namely the pressure in the oil phase, p, which is computed implicitly. We remark that β_2 and β_3 in $\tilde{T}$ are regarded as constants relative to the difference operator Δ when evaluating the left-hand side; i.e., they are taken outside of the operator.

As in the two-phase incompressible case, Eq. 7.64 leads to a banded matrix problem (tridiagonal for 1-D, pentadiagonal for 2-D, and heptadiagonal for 3-D). In fact, this structure is typical of reservoir simulators, even for other formulations. An IMPES formulation is characterized by single elements in each entry of the matrix whereas the simultaneous formulation has entries which are in fact $r \times r$ submatrices (for r unknowns) as will be subsequently seen. Consequently, solution of the matrix problem for an IMPES model entails slightly more work (in coefficient generation) than that of a single-phase flow simulator. Of course additional solution time is encountered in performing the subsequent saturation computations. Nevertheless, IMPES models are the cheapest to run and provide adequate results in many instances.

If we use the Douglas-Rachford procedure, to solve Eq. 7.64 then the expansions are

x-sweep:

$$\Delta_x \tilde{T}_x \Delta_x p^{k+1/3} + \Delta_y \tilde{T}_y \Delta_y p^k + \Delta_z \tilde{T}_p \Delta_z p^k = \beta(p^{k+1/3} - p^n)$$
$$+ (\Sigma \tilde{T})\sigma_k(p^{k+1/3} - p^k) + G \quad (7.65)$$

y-sweep:

$$\Delta_x \tilde{T}_x \Delta_x p^{k+1/3} + \Delta_y \tilde{T}_y \Delta_y p^{k+2/3} + \Delta_z \tilde{T}_z \Delta_z p^k = \beta(p^{k+2/3} - p^n)$$
$$+ (\Sigma \tilde{T})\sigma_k(p^{k+2/3} - p^k) + G \quad (7.66)$$

z-sweep:

$$\Delta_x \tilde{T}_x \Delta_x p^{k+1/3} + \Delta_y \tilde{T}_y \Delta_y p^{k+2/3} + \Delta_z \tilde{T}_z \Delta_z p^{k+1} = \beta(p^{k+1} - p^n)$$
$$+ (\Sigma \tilde{T})\sigma_k(p^{k+1} - p^k) + G \quad (7.67)$$

where

$$\Sigma \tilde{T} = \sum_{l=1}^{3} (T_{l\,i+1/2jk} + T_{l\,i-1/2jk} + T_{l\,ij+1/2k} + T_{l\,ij-1/2k} + T_{l\,ijk+1/2} + T_{l\,ijk-1/2}).$$

Using steps similar to those employed before, Eqs. 7.65–7.67 can be put in the form

$$x: \quad \Delta_x \tilde{T}_x \Delta_x PX - (\beta + \sigma_k \Sigma \tilde{T}) PX = -\tilde{G} \quad (7.68)$$

$$y: \quad \Delta_y \tilde{T}_y \Delta_y PY - (\beta + \sigma_k \Sigma \tilde{T}) PY = -(\beta + \sigma_k \Sigma \tilde{T}) PX \quad (7.69)$$

$$z: \quad \Delta_z \tilde{T}_z \Delta_z PZ - (\beta + \sigma_k \Sigma \tilde{T}) PZ = -(\beta + \sigma_k \Sigma \tilde{T}) PY \quad (7.70)$$

where PX, PY and PZ are given in Eqs. 7.22a–7.22c with p substituting for Φ_w and $\tilde{G} = \Delta \tilde{T} \Delta p - \beta(p^k - p^n) - G$. At the end of each iteration the pressures are updated using $p^{k+1} = p^k + PZ$. Convergence is achieved when

$$\frac{\Sigma |\tilde{G}|}{\Sigma |q_w|} \leq \epsilon.$$

After solving for p^{n+1}, the water and oil saturations are computed directly using Eqs. 7.59 and 7.60 in a manner analogous to that outlined for the two-phase flow problem. The gas saturation S_g^{n+1} is computed by difference. S_w^{n+1} and S_o^{n+1} are also used to compute new capillary pressures at time level $n+1$. This completes the calculations over a time step.

7.3.2 Stability of an IMPES Formulation

For problems involving rapid changes in pressure and saturation (such as well coning) the IMPES formulation becomes unstable for practical time-step sizes. To convey the gist of the time-step limitations, we linearize Eqs. 7.44–7.46 by invoking some simplifying assumptions. A nonlinear stability analysis involves considerable mathematical development[8] and is not required for our purposes. The assumptions are: (1) flow is incompressible; (2) no phase transfers occur; (3) the transmissibilities are constant and uniform; (4) the gravity effects are negligible. Again, we use a development due to Coats.[7] Eqs. 7.44–7.46 then become

$$T_w \Delta^2 p_w + B_w q_w = \frac{V_p}{\Delta t} \Delta_t S_w \qquad (7.71)$$

$$T_o \Delta^2 p_o + B_o q_o = \frac{V_p}{\Delta t} \Delta_t S_o \qquad (7.72)$$

$$T_g \Delta^2 p_g + B_g q_g = \frac{V_p}{\Delta t} \Delta_t S_g \qquad (7.73)$$

Using the capillary-pressure relationships given in Eqs. 7.48 and 7.49 we have

$$\Delta p_w = \Delta p_o - P'_{cwo} \Delta S_w$$

$$\Delta p_g = \Delta p_o + P'_{cgo} \Delta S_g$$

where

$$P'_{cwo} = \frac{dP_{cwo}}{dS_w}$$

$$P'_{cgo} = \frac{dP_{cgo}}{dS_g}$$

Assuming that P'_{cwo} and P'_{cgo} are linear for every S_w and S_g we have

$$\Delta^2 p_w = \Delta^2 p_o - P'_{cwo} \Delta^2 S_w \qquad (7.74)$$

$$\Delta^2 p_g = \Delta^2 p_o + P'_{cgo} \Delta^2 S_g \qquad (7.75)$$

Substitution of Eqs. 7.74 and 7.75 in Eqs. 7.71 and 7.73 yields

$$T_w(\Delta^2 p - P'_{cwo} \Delta^2 S_w) + B_w q_w = \frac{V_p}{\Delta t} \Delta_t S_w \qquad (7.76)$$

$$T_o \Delta^2 p + B_o q_o = \frac{V_p}{\Delta t} \Delta_t S_w - \frac{V_p}{\Delta t} \Delta_t S_g \qquad (7.77)$$

$$T_g(\Delta^2 p + P'_{cgo} \Delta^2 S_g) + B_g q_g = \frac{V_p}{\Delta t} \Delta_t S_g \qquad (7.78)$$

where $\Delta^2 p_o \equiv \Delta^2 p$.

For the three unknowns S_w, S_g, and p assume errors are introduced with magnitudes ϵ_1, ϵ_2, ϵ_3, respectively, that satisfy Eqs. 7.76–7.78. Consequently,

$$T_w \Delta^2 \epsilon_3^{n+1} + \tilde{T}_w \Delta^2 \epsilon_1^n = \frac{V_p}{\Delta t} \Delta_t \epsilon_1 \qquad (7.79)$$

$$T_o \Delta^2 \epsilon_3^{n+1} = -\frac{V_p}{\Delta t} \Delta_t \epsilon_1 - \frac{V_p}{\Delta t} \Delta_t \epsilon_2 \qquad (7.80)$$

$$T_g \Delta^2 \epsilon_3^{n+1} + \tilde{T}_g \Delta^2 \epsilon_2^n = \frac{V_p}{\Delta t} \Delta_t \epsilon_2 \qquad (7.81)$$

where $\tilde{T}_w \equiv -T_w P'_{cwo}$ and $\tilde{T}_g \equiv T_g P'_{cgo}$.

The error component is given by

$$\epsilon_{ijk}^n = \zeta^n e^{J(i\alpha_x + j\alpha_y + k\alpha_z)}; \quad J \equiv \sqrt{-1}. \qquad (7.82)$$

Therefore,

$$\Delta_x^2 \epsilon_{ijk}^n = -\gamma_x \epsilon_{ijk}^n \qquad (7.83)$$

$$\Delta_y^2 \epsilon_{ijk}^n = -\gamma_y \epsilon_{ijk}^n \qquad (7.84)$$

$$\Delta_z^2 \epsilon_{ijk}^n = -\gamma_z \epsilon_{ijk}^n \qquad (7.85)$$

where

Multiphase Flow

$$\gamma_x = 4 \sin^2\left(\frac{\alpha_x}{2}\right)$$

$$\gamma_y = 4 \sin^2\left(\frac{\alpha_y}{2}\right)$$

$$\gamma_z = 4 \sin^2\left(\frac{\alpha_z}{2}\right)$$

Thus,

$$T_l \Delta^2 \epsilon_m = T_{lx} \Delta_x^2 \epsilon_m + T_{ly} \Delta_y^2 \epsilon_m + T_{lz} \Delta_z^2 \epsilon_m = -\lambda_l \gamma_m \epsilon_m, \ l = w, o, g; \ m = 1, 2, 3$$

where $\lambda_l = k_{rl}/\mu_l$ and

$$\gamma_m = \frac{4 \, k_x \Delta y \Delta z}{\Delta x} \sin^2\left(\frac{\alpha_{mx}}{2}\right) + \frac{4 \, k_y \Delta x \Delta z}{\Delta y} \sin^2\left(\frac{\alpha_{my}}{2}\right) + \frac{4 \, k_z \Delta x \Delta y}{\Delta z} \sin^2\left(\frac{\alpha_{mz}}{2}\right).$$

Consequently, the error equations become

$$-\lambda_w \gamma_3 \epsilon_3^{n+1} - \tilde{\lambda}_w \gamma_1 \epsilon_1^n = \beta(\epsilon_1^{n+1} - \epsilon_1^n) \tag{7.86}$$

$$-\lambda_o \gamma_3 \epsilon_3^{n+1} = -\beta(\epsilon_1^{n+1} - \epsilon_1^n) - \beta(\epsilon_2^{n+1} - \epsilon_2^n) \tag{7.87}$$

$$-\lambda_g \gamma_3 \epsilon_3^{n+1} - \tilde{\lambda}_g \gamma_2 \epsilon_2^n = \beta(\epsilon_2^{n+1} - \epsilon_2^n) \tag{7.88}$$

where $\tilde{\lambda}_w = -P'_{cwo}\lambda_w$, $\tilde{\lambda}_g = -P'_{cgo}\lambda_g$, and $\beta = V_p/\Delta t$.

Eqs. 7.86–7.88 are then added to get

$$\lambda_w \gamma_3 \epsilon_3^{n+1} + \tilde{\lambda}_w \gamma_1 \epsilon_1^n + \lambda_o \gamma_3 \epsilon_3^{n+1} + \lambda_g \gamma_3 \epsilon_3^{n+1} + \tilde{\lambda}_g \gamma_2 \epsilon_2^n = 0. \tag{7.89}$$

If we solve Eq. 7.89 for $\gamma_3 \epsilon_3^{n+1}$ and substitute the result in Eqs. 7.86 and 7.87, then

$$\beta \epsilon_1^{n+1} = \left[\beta - \frac{\tilde{\lambda}_w}{\lambda}(\lambda_o + \lambda_g)\gamma_1\right] \epsilon_1^n + \left(\frac{\lambda_w \tilde{\lambda}_g}{\lambda}\right) \gamma_2 \epsilon_2^n \tag{7.90}$$

$$\beta \epsilon_1^{n+1} + \beta \epsilon_2^{n+1} = \left[\beta - \frac{\lambda_o \tilde{\lambda}_w}{\lambda} \gamma_1\right] \epsilon_1^n + \left[\beta - \frac{\lambda_o \tilde{\lambda}_g}{\lambda} \gamma_2\right] \epsilon_2^n \tag{7.91}$$

where $\lambda \equiv \lambda_o + \lambda_g + \lambda_w$. These can be written in matrix form as

$$\mathbf{A}\mathbf{e}^{n+1} = \mathbf{B}\mathbf{e}^n \text{ or } \mathbf{e}^{n+1} = \mathbf{A}^{-1}\mathbf{B}\mathbf{e}^n = \mathbf{C}\mathbf{e}^n \tag{7.92}$$

where

$$C = \frac{1}{\beta} \begin{bmatrix} \beta - \gamma_1 \tilde{\lambda}_w \left(1 - \frac{\lambda_w}{\lambda}\right) & \frac{\lambda_w \tilde{\lambda}_g \gamma_2}{\lambda} \\ \frac{\tilde{\lambda}_w \lambda_g \gamma_1}{\lambda} & \frac{\tilde{\lambda}_g \gamma_2 (\lambda_o + \lambda_w)}{\lambda} \end{bmatrix}$$

For stability and convergence we require that $|\rho(C)| < 1$ or for any eigenvalue σ, we require $-1 < \sigma < 1$. The characteristic equation is

$$\begin{vmatrix} C_{11} - \sigma & C_{12} \\ C_{21} & C_{22} - \sigma \end{vmatrix} = 0.$$

Thus,

$$\sigma_{1,2} = \frac{-(C_{11} + C_{22}) \pm \sqrt{(C_{11} + C_{22})^2 - 4\psi_c}}{2} \tag{7.93}$$

where $\psi_c \equiv C_{11} C_{22} - C_{12} C_{21}$.

Substituting the definitions of the C's we get

$$\sigma_{max} = \frac{1}{2\beta} \left\{ 2\beta - \frac{\gamma_1 \tilde{\lambda}_w (\lambda_o + \lambda_g)}{\lambda} - \frac{\gamma_2 \tilde{\lambda}_g (\lambda_o + \lambda_w)}{\lambda} - \frac{1}{\lambda} [\gamma_1 \tilde{\lambda}_w (\lambda_o + \lambda_g) \right.$$
$$\left. - \gamma_2 \tilde{\lambda}_g (\lambda_o + \lambda_w)^2 + 4 \lambda_w \lambda_g \tilde{\lambda}_w \tilde{\lambda}_g \gamma_1 \gamma_2]^{1/2} \right\}. \tag{7.94}$$

This implies that $\sigma_{max} < 1$. However, for $\sigma_{max} > -1$, we get the following restriction

$$\Delta t \leq \frac{2 V_p}{(2\beta - \alpha)} \tag{7.95}$$

where α represents the last three terms in the curly brackets in Eq. 7.94. Thus we have conditional stability for the IMPES method. For a compressible system, however, this restriction may be relaxed somewhat. The conditional stability arises from explicit treatment of capillary pressure and the transmissibilities.

The restriction given by Eq. 7.95 is not important in areal (x,y) calculations where the grid block dimensions are roughly equal. It becomes important in three-dimensional calculations and cross-sectional calculations where Δz is small relative to Δx and Δy. For regions where gas and oil are flowing, then $\lambda_w = \tilde{\lambda}_w = 0$, and the inequality in Eq. 7.95 becomes

$$\Delta t \leq \frac{\phi \Delta x \Delta y \Delta z \left(\frac{\mu_o}{k_{ro}} + \frac{\mu_g}{k_{rg}}\right)_{min}}{2 P'_{cgo} \left(k_x \frac{\Delta y \Delta z}{\Delta x} + k_y \frac{\Delta x \Delta z}{\Delta y} + k_z \frac{\Delta x \Delta y}{\Delta z}\right)}. \tag{7.96}$$

Multiphase Flow

This is the conditional stability restriction caused by explicit treatment of the P_c's. For regions where water and oil are flowing, we get an identical expression to Eq. 7.96 except μ_g/k_{rg} is replaced by μ_w/k_{rw}, and P'_{cgo} by P'_{cwo}. In both cases the time-step restrictions caused by explicit transmissibilities are not reflected. For areal calculations $k_z = 0$, and if $\Delta x \approx \Delta y$ and $k_x = k_y$ then we get

$$\Delta t \leq \min \left\{ \frac{\phi \Delta x \Delta y \left(\frac{\mu_o}{k_{ro}} + \frac{\mu_g}{k_{rg}}\right)_{\min}}{4 k P'_{cgo}}, \frac{\phi \Delta x \Delta y \left(\frac{\mu_o}{k_{ro}} + \frac{\mu_w}{k_{rw}}\right)_{\min}}{4 k P'_{cwo}} \right\}. \quad (7.97)$$

The ratio of the maximum time-step size in cross-sectional or three-dimensional studies to the maximum time-step size in two-dimensional studies is then

$$\frac{\Delta t_{3\text{-D}}}{\Delta t_{2\text{-D}}} = \frac{2 k \Delta z^2}{k_z \Delta x \Delta y}. \quad (7.98)$$

This is usually of the order of 0.0010.

The stability restriction imposed by explicit treatment of the transmissibilities is dependent primarily on the fractional flow of each phase. If we assume that the fractional flow of a phase is dependent only on the saturation of that phase, then we define

$$f(S_l) = \frac{\lambda_l}{\lambda} \quad l = o, w, g. \quad (7.99)$$

If we assume zero P_c's and zero compressibility we get the restriction

$$\Delta t \leq \min \left\{ \frac{\phi}{f'_l \left(\frac{U_x}{\Delta x} + \frac{U_y}{\Delta y} + \frac{U_z}{\Delta z}\right)} \right\}_l \quad l = o, w, g \quad (7.100)$$

where U_x, U_y, U_z are the total Darcy velocities in each of the three directions and

$$f'_l = \frac{df(S_l)}{dS_l}. \quad (7.101)$$

The IMPES method with explicit transmissibilities and P_c's will handle most reservoir problems. However, in problems involving highly unfavourable mobility ratios or rapidly converging flow, some difficulty may be encountered with such a model. This can usually be overcome by computing the transmissibilities and/or capillary pressures implicitly or semi-implicitly. Blair and Weinaug[9] describe a completely implicit approach for a well-bore coning model. MacDonald and Coats[10] and later Nolen and Berry[11] give a treatment using semi-implicit approaches to the same problem. The Blair-Weinaug approach greatly relaxes the time-step restrictions but it requires considerable computational work at each time step. Essentially the same results can be obtained with less work by expressing the nonlinear coefficients semi-implicitly. Furthermore, no significant time-step restrictions are realized by this approach. We briefly illustrate the technique with regard to transmissibility.

In this formulation we use a Taylor's series expansion and neglect terms of order two and higher. Thus,

$$T_l^{n+1} \approx T_l^n + \left(\frac{dT_l}{dk_{rl}}\right)^n \left(\frac{dk_{rl}}{dS_l}\right) \Delta S_l + O(\Delta S_l^2) \qquad (7.102)$$

where

$$\Delta S_l = S_l^{n+1} - S_l^n, \qquad (7.103)$$

dT_l^n/dk_{rl} is evaluated analytically, and dk_{rl}/dS_l is either evaluated at time level n as done by MacDonald and Coats[10] or as a chord slope (Nolen and Berry[11]), i.e.,

$$\frac{dk_{rl}}{dS_l} = \frac{k_{rl}(S_l^n + \delta S_l) - k_{rl}(S_l^n)}{\delta S_l} \qquad (7.104)$$

where δS_l is an increment in saturation that is specified arbitrarily. If we substitute Eq. 7.102 in the flow equations for the IMPES method we get terms of the form

$$(T_l^{n+1})_{i+1/2} (\Phi_{i+1}^{n+1} - \Phi_i^{n+1}) = [(T_l^n)_{i+1/2} + \Delta T_l](\Phi_{i+1}^{n+1} - \Phi_i^{n+1})$$
$$= [(T_l^n)_{i+1/2} + \Delta T_l](\Phi_{i+1}^n - \Phi_i^n) + [(T_{l\,i+1/2}^n + \Delta T_l)(\Delta\Phi_{i+1} - \Delta\Phi_i)] \qquad (7.105)$$

$$\approx [(T_l^n)_{i+1/2} + \Delta T_l](\Phi_{i+1}^n - \Phi_i^n) + (T_l^n)_{i+1/2}(\Delta\Phi_{i+1} - \Delta\Phi_i) \qquad (7.106)$$

where $\Delta T_l = (dT_l/dk_{rl})/(dk_{rl}/dS_l \, \Delta S_l)$ and we have neglected terms of the form $\Delta T_l \Delta \Phi_l$ to give a better linearization. If the transmissibilities are updated at each iteration level, then the transmissibilities at the converged iteration level will approximate those at time level $n + 1$. For most time steps normally used, i.e., $\Delta t \leq 90$ days, this linearization does not produce significant truncation errors. Similar techniques are used for computing P_c^{n+1} and semi-implicit rates, q_l^{n+1}.

Another technique called the *sequential method* provides for an implicit computation of saturation.[12] The procedure involves two steps. First p^{n+1} is computed from Eq. 7.64 with semi-implicit rates and transmissibilities. This pressure is employed in a reformulation of Eqs. 7.59–7.61, where p^{n+1} is now a known, to compute two of the saturations implicitly. The second step treats flow as incompressible. Moreover, it does not preserve material balances in all phases and various artifices must be employed to correct for this. The work involved is approximately twice that of IMPES. Its principal advantage is more stability than IMPES and less computer effort than a fully implicit procedure.

7.3.3 Fully Implicit Formulation

The fully implicit formulation provides for a simultaneous solution of three unknowns. Its chief advantage is unconditional stability. However, the computer work required to obtain a solution is roughly seven times that of an IMPES model. Nevertheless, fully implicit models are widely used in black-oil systems for coning studies and, at least in one instance, for multiwell simulations.[13] The auxiliary relationships in Eqs. 7.47–7.49 are used to eliminate three of the six unknowns (p_l and S_l, $l = w, o, g$) in Eqs. 7.44–7.46. Typically the three remaining unknowns are the phase potentials (or pressures) or two saturations and a phase pressure. If three potentials are the unknowns, then nonzero capillary pressure data are required, in particular nonzero slopes of P_c vs. saturation. Should one wish to simulate a situation where they are zero, a fictitious P_c-slope can be introduced which is sufficiently small to yield results essentially identical to the zero-slope case. The chief advantage of this formulation is that the unknowns are smoothly varying in spite of saturation discontinuities. This leads to more rapid convergence when iterative procedures are employed for solving the matrix problem. For direct methods, the requirement of nonzero capillary data is an unnecessary limitation and the alternate formulation is preferred. In this case, the three unknowns are typically p_o, S_w, and S_g. With wider usage of direct methods, the trend today is to use the alternate formulation. For our purpose here, we assume this case and sketch just the essentials.

To distinguish between changes over an iteration and changes over a time step, we employ the operator δx to mean the same as Δ_t and δ to indicate a change from iteration level k to $k + 1$. Thus, $\bar{\delta} x = x^{n+1} - x^n$ and $\delta x = x^{k+1} - x^k$. The relationship between the two is $\bar{\delta} x \approx x^{k+1} - x^n = x^k + \delta x - x^n$. The fully implicit formulation of Eqs. 7.44–7.46 can be written in the compact form

$$\Delta T_l^{n+1} \Delta \Phi_l^{n+1} + q_l^{n+1} + \omega \{\Delta (T_o R_s)^{n+1} \Delta \phi_o^{n+1} + (q_o R_s)^{n+1}\}$$
$$= \frac{V_b}{\Delta t} \bar{\delta} \{\phi b_l S_l + \omega(\phi b_o R_s S_o)\}, \quad l = w, o, g \quad (7.107)$$

where

$$\omega \equiv \begin{cases} 1, \ell = g \\ 0, \ell = w, o. \end{cases}$$

Writing this expression in terms of the δ-operator gives

$$\Delta(T_\ell^k + \delta T_o)\Delta(\Phi_\ell^k + \delta\Phi_\ell) + q_\ell^k + \delta q_\ell$$
$$+ \omega\{\Delta[(T_oR_s)^k + \delta(T_oR_s)]\Delta(\Phi_o^k + \delta\Phi_o) + (q_oR_s)^k + \delta(q_oR_s)\}$$
$$= \frac{V_b}{\Delta t}\{[\phi b_\ell S_\ell + \omega(\phi b_o R_s S_o)]^k + \delta[\phi b_\ell S_\ell + \omega(\phi b_o R_s S_o)]$$
$$- [\phi b_\ell S_\ell + \omega(\phi b_o R_s S_o)]^n\} \quad \ell = w, o, g. \quad (7.108)$$

In the expansion of Eq. 7.108, we drop second order terms of the form $\Delta(\delta x)\Delta(\delta y)$. The residual, at the kth iterate is given by

$$R_\ell^k \equiv \Delta T_\ell^k \Delta \Phi_\ell^k + q_\ell^k + \omega\{\Delta(T_oR_s)^k\Delta\Phi_o^k + (q_oR_s)^k\}$$
$$- \frac{V_b}{\Delta t}\{[\phi b_\ell S_\ell + \omega(\phi b_o R_s S_o)]^k - [\phi b_\ell S_\ell + \omega(\phi b_o R_s S_o)]^n\}$$
$$\ell = w, o, g. \quad (7.109)$$

Consequently, Eq. 7.108 can be written in the *residual form*

$$\Delta(\delta T_\ell)\Delta\Phi_\ell^k + \Delta T_\ell^k\Delta(\delta\Phi_\ell) + \delta q_\ell$$
$$+ \omega\{\Delta\delta(T_oR_s)\Delta\Phi_o^\ell + \Delta(T_oR_s)^\ell\Delta(\delta\Phi_o) + \delta(q_oR_s)\}$$
$$= \frac{V_b}{\Delta t}\delta\{\phi b_\ell S_\ell + \omega(\phi b_o R_s S_o)\} - R_\ell^k$$
$$\ell = w, o, g. \quad (7.110)$$

For a convergent process, $R_\ell^k \to 0$ as $k = 1, 2, \ldots$ for $\ell = w, o, g$. Instead of expressing porosity in terms of p_w as in Eq. 7.50, we express it in terms of p_o which is considered synonymous with reservoir pressure, p. Thus, $\phi = \phi_o(1 + c_r p)$. The right-hand sides of Eq. 7.110 are expanded in terms of the three primary unknowns, δp, δS_w, and δS_g where again a consistent expansion is

$$\delta(ab) = a^{k+1}\delta b + b^k \delta a. \quad (7.111)$$

Each expansion has the form

$$RHS_\ell = C_{\ell 1}\delta p + C_{\ell 2}\delta S_w + C_{\ell 3}\delta S_g - R_\ell^k \quad (7.112)$$

where here index $\ell = 1, 2, 3$ corresponding to the water, oil, and gas equations. Thus, for the water equation:

where

$$D_{21} = WI\lambda_o \equiv PI_o \qquad (7.143)$$

$$D_{22} = WI\frac{b_o}{\mu_o} k'_{row}(p - p_{wf}) \qquad (7.144)$$

$$D_{23} = WI\frac{b_o}{\mu_o} k'_{rog}(p - p_{wf}) \qquad (7.145)$$

In a similar manner we evaluate $\delta(q_g)$, $\delta(q_w)$, and $\delta(R_s q_o)$ to generate a set of D-coefficients which makes up the elements of a *well matrix*, **D**. The actual form these coefficients take depends upon the specifications on a well. This is treated in more detail in the next chapter.

In matrix form Eq. 7.125 becomes

$$\mathbf{TX + DX = CX + R} \qquad (7.146)$$

or simply

$$\mathbf{AX = R} \qquad (7.147)$$

where $\mathbf{A \equiv T + D - C}$ and **R** is a vector containing the kth iteration level residuals. Matrix **A** is a band matrix identical in structure to those previously discussed. However, each entry (e.g., the x's in Figs. 6.6 and 6.7) is a 3×3 submatrix corresponding to each active grid block in the reservoir. The first, second and third rows within a submatrix contain the coefficients for δp, δS_w, and δS_g (in that order) that come from the water, oil, and gas equations, respectively. If one or more of these phases is not flowing in a particular grid block then some of the elements within the corresponding submatrix are zero. However, the main diagonal of **A** is always nonzero since it contains the **D**- and **C**-coefficients and darcian flow terms from **T**. If we consider a two-phase flow problem, then **A** consists of 2×2 submatrices arranged in band form. Obviously single phase flow and the matrix problem in Eq. 7.64 constitute the degenerate case where the submatrices reduce to a single element.

The matrix problem in Eq. 7.147 can be solved by direct Gaussian elimination, SIP or the block SOR methods discussed in chapter 6. For Gaussian elimination, alternate diagonal ordering is still applicable. For 3-D problems, it can be extended to alternate diagonal planes.[14] For large problems, this approach loses its advantage, and an iterative procedure is advisable to minimize storage. One can use LSOR or 2LSOR or, alternatively, the multiphase flow algorithm for SIP.[15] Solution of Eq. 7.147 produces the change in the primary variables over an iteration (the *outer itera-*

tion if an iterative technique is used to solve the matrix problem). Usually 2 to 4 iterations are required to complete the time step.

7.3.4 The Adaptive Implicit Method

There are obvious advantages to both an IMPES simulator and a fully implicit one. The former provides adequate results at minimum cost in many instances. However, it can be inadequate for problems where rapid surges in the dependent variables occur. These problems generally require an implicit calculation for those variables experiencing such surges to maintain stability. A fully implicit simulator provides such capability. However, the additional computer expense is substantial. Furthermore, time truncation errors are generally higher in implicit simulators.[16] It is clear that rapid changes in pressure and saturation don't occur over an entire reservoir simultaneously. Rather this is a local phenomenon usually restricted to the proximity of wells or aquifer-reservoir boundaries. For example, consider three wells penetrating the same reservoir which are sufficiently distant that interference effects are minimal. Suppose one well is producing at such a high rate so as to locally distort the gas-oil and water-oil contacts. Another, producing at moderate rate, is perforated at a lower elevation in the oil zone causing a distortion only in the water-oil contact. The third, a low rate well, has a producing interval far removed from the fluid contacts. The first well will normally require an implicit computation of p, S_w, S_g in those grid blocks affected by the fluid contact distortions. In the blocks surrounding the second well, an implicit calculation of p and S_w would be sufficient. And in the third, one need treat only p implicitly. Moreover, in the intervening blocks, an implicit calculation of p (an IMPES model) would probably be sufficient.

A fully implicit model can adequately handle the situation described above. Clearly, however, there would be a large amount of overkill in many of the grid blocks, i.e., more implicitness is supplied than is required. On the other hand, an IMPES model would lead to underkill and instabilities for practical time step sizes in those grid blocks requiring more implicitness. Unfortunately, simulators offering a fixed level of implicitness do not recognize that only a small fraction of the total number of grid blocks experience sufficiently large surges in pressures and/or saturation to justify implicit treatment. When it is needed, it may not be required in all phases, or for long periods of time. Moreover, those cells requiring implicit calculations will change as the simulation advances.

Consideration of this problem has led to the recent development of the adaptive implicit method (AIM).[17,18] Because some confusion exists in the technical literature with respect to the exact meaning of *fully implicit*,[16] or *strongly coupled*,[13] we prefer to speak of *degrees of implicitness*.† We identify degrees of implicitness with the number of dependent variables, computed implicitly. Thus an IMPES calculation is first degree and a fully implicit calculation is third degree. With AIM, various degrees of implicitness are invoked regionally or individually cell-by-cell, i.e. the solution is advanced with adjacent cells having different degrees of implicitness. As the calculations proceed, the degrees of implicitness locally and dynamically shift as needed—all automatically. The idea is to apply the right amount of implicitness where and when needed and only as long as needed. The advantages of this approach are obvious. First, the user need not make a choice regarding an IMPES model vs. a fully implicit one. There is also no problem of over/under-kill. Thus, in our hypothetical three well reservoir, third, second, and first degree calculations would be performed locally around the first, second, and third wells, respectively, while the intervening blocks are first degree. Finally, one achieves an optimum with regard to the stability-truncation error-cost question.

In the adaptive implicit method, the order of matrix **A** in Eq. 7.147 is not fixed but is permitted to change with time. The order at any moment is determined by the total number of variables to be computed implicitly. The shift in order can occur within a time step since Eq. 7.147 is based on changes over an iteration. The shift is accomplished by performing the *equivalent* of elementary row operations in **A** and then setting certain coefficients to zero.

7.4 Exercises

1. Consider the two adjacent grid blocks below in which flow of oil is from left to right, i.e., $(\Phi_o)_i > (\Phi_o)_{i+1}$.

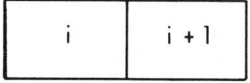

   Suppose S_o in block i is at or close to the irreducible oil saturation, S_{or}, while in block $i+1$, $S_o \gg S_{or}$. What is the consequence if arithmetic averaging is employed to determine $(k_{ro})_{2+1/2}$?

2. Devise a scheme for determining which block is the upstream block in a simulator.

---

† An essential common element in the terminology, "fully implicit" and "strongly coupled" is that all dependent variables are computed implicitly. Some differences in meaning between various authors occurs regarding the level of implicit treatment of transmissibilities, rates, etc.

3. Show that if the Douglas-Rachford iterative ADI procedure is applied in an IMPES model, then Eqs. 7.65–7.67 can be put in the form given by Eqs. 7.68–7.70.
4. Show that the matrix structure for an IMPES treatment of three-phase flow in the 2-D reservoir in Fig. 6.4 is identical to that given in Fig. 6.7 and that each entry in the matrix is a single element.
5. Suppose if instead of Eq. 7.141 we use the expansion

$$\delta(Zq_l) = Z^{k+1}\delta(q_l) + q_l^k \delta Z$$

$$= Z^{k+1}\frac{WI}{\mu_l}\delta(k_{rl}b_l\Delta p) + q_l^k \delta Z$$

$$= Z^{k+1}\frac{WI}{\mu_l}\{(k_{rl}\Delta p)^k \delta b_l + b_l^{k+1}[k_{rl}^k \delta(\Delta p) + \Delta p^{k+1}\delta k_{rl}]\} + q_l^k \delta Z.$$

Would the D-coefficients in Eqs. 7.143–7.145 be different for an oil producing well with constant p_{wf}?

6. Construct the structure of the matrix problem in Eq. 7.147 for a 1-D reservoir having four grid blocks. Show the elements of **A**, **X**, and **b**, and partition each to show the submatrices. What solution technique would you use for this problem?

7.5 References

1. Coats, K.H., and Nielson, R.L., Terhune, M.H., and Weber, A.H.: "Simulation of Three-Dimensional, Two-Phase Flow in Oil and Gas Reservoirs," *Soc. Pet. Eng. J.* (Dec. 1967) 377.
2. Peaceman, D.W.: "Numerical Solution of the Nonlinear Equations for Two-Phase Flow Through Porous Media," *Nonlinear Partial Differential Equations*, Academic Press, New York City (1967).
3. Todd, M.R., O'dell, P.M., and Hirasaki, G.J.: "Methods for Increased Accuracy in Reservoir Simulators," *Trans.*, AIME (1972) **253**, 515.
4. Douglas, J., Peaceman, D.W., and Rachford, H.H.: "A Method for Calculating Multi-Dimensional Immiscible Displacement," *Trans.*, AIME (1959) **216**, 297.
5. Sheldon, J.W., Harris, C.D., and Bavly, D.: "A Method for General Reservoir Behavior Simulation on Digital Computers," paper SPE 1521-G presented at the SPE 35th Annual Meeting, Denver, Oct. 2–5, 1960.
6. Stone, H.L. and Garder, A.O., Jr.: "Analysis of Gas-Cap or Dissolved Gas Drive Reservoirs," *Trans.*, AIME (1961) **222**, 92.
7. Coats, K.H.: *Elements of Hydrocarbon Reservoir Simulation*, Intercomp. Resource Development & Engineering, Inc., Houston (1972).
8. Peaceman, D.W.: "A Nonlinear Stability Analysis for Difference Equations Using Semi-Implicit Mobility," paper SPE 5735 presented at the 4th Symposium on Numerical Simulation of Reservoir Performance, Los Angeles, Feb. 19–20, 1976.
9. Blair, P.M. and Weinaug, C.F.: "Solution of Two-Phase Flow Problems Using Implicit Difference Equations," *Trans.*, AIME (1969) **246**, 417.
10. MacDonald, R.C. and Coats, K.H.: "Methods for Numerical Simulation of Water and Gas Coning," *Trans.*, AIME (1970) **249**, 425.
11. Nolen, J.S. and Berry, D.W.: "Tests of the Stability and Time-Step Sensitivity of Semi-Implicit Reservoir Simulation Techniques," *Trans.*, AIME (1972) **253**, 253.

12. Spillette, A.G., Hillestad, J.G., and Stone, H.L.: "A High-Stability Sequential Solution Approach to Reservoir Simulation," paper SPE 4542 presented at the SPE 48th Annual Meeting, Las Vegas, Sept. 30–Oct. 3, 1973.
13. Bansal, P.P., Harper, J.L., McDonald, A.E., Moreland, E.E., Odeh, A.S., and Trimble, R.H.: "A Strongly Coupled, Fully Implicit, Three Dimensional, Three Phase Reservoir Simulator," paper SPE 8329 presented at the SPE 54th Annual Technical Conference and Exhibition, Las Vegas, Sept. 23–26, 1979.
14. Price, H.S. and Coats, K.H.: "Direct Methods in Reservoir Simulation," *Trans.*, AIME (1974) **257**, 295.
15. Weinstein, H.G., Stone, H.L., and Kwan, T.V.: "Simultaneous Solution of Multiphase Reservoir Flow Problems," *Soc. Pet. Eng. J.* (June 1970) 99.
16. Settari, A. and Aziz, K.: "Treatment of Nonlinear Terms in the Numerical Solution of Partial Differential Equations for Multiphase Flow in Porous Media," *Int. J. Multiphase Flow* (April 1975) **1**, No. 16, 817–844.
17. Thomas, G.W. and Thurnau, D.H.: "Reservoir Simulation using an Adaptive Implicit Method," paper SPE 10120 to be presented at the SPE 1981 Fall meeting.
18. Thomas, G.W. and Thurnau, D.H.: "The Mathematical Basis of the Adaptive Implicit Method," forthcoming.

8
Special Concepts

*Although the whole of this life were said
to be nothing but a dream and the physical
world nothing but a phantasm, I
should call this dream or phantasm
real enough, if, using reason well,
we were never deceived by it.*

Leibniz

8.1 Introduction

Thus far we have set forth the major aspects of a reservoir simulator. In this chapter, attention is directed to special features that a simulator may provide. We include here a discussion of how wells are treated. This is regarded in the context of a special concept since our concerns to now have been with darcian flow through global porous systems. Flow into or from a well is a local phenomenon and certain aspects of it are non-darcian. Moreover, wells constitute the control points in a reservoir, and special concepts must be invoked to simulate these controls. Included here is a treatment of pseudo functions which are sometimes employed primarily to reduce dimensionality of a simulation problem. Attention is also given to how gas percolation is controlled in some IMPES models that employ explicit transmissibilities. The chapter concludes with some brief remarks regarding variable bubble-point problems.

8.2 Treatment of Wells

In the finite difference representation of the flow equations, we indicated the presence of source/sink terms by q_l, $l = o, w, g$. These represent produc-

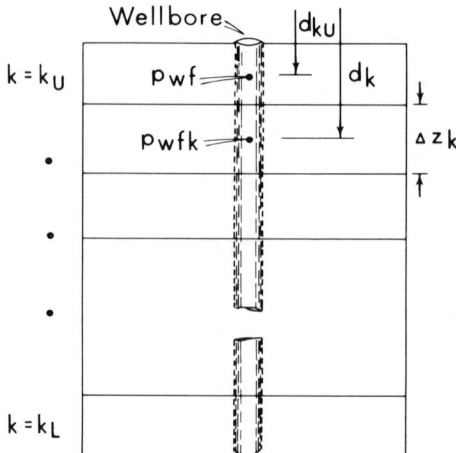

Fig. 8.1 Wellbore Configuration.

tion or injection for the entire grid block in which they occur. If there is more than one well in a single block, then q_l represents the algebraic sum of all injection and production for the lth phase in the block. If a well is completed in more than one grid block in the vertical direction, the total well rate is allocated (as sources and sinks) to each of these well blocks. Consider the situation depicted in Fig. 8.1 where a well is presumed completed between block k_U (upper) and block k_L (lower).†

For layer k,

$$q_{lk} = WI_k (p_k - p_{wfk}) \lambda_l \tag{8.1}$$

where the wellbore index is given by

$$WI_k = \alpha \left\{ \frac{k \Delta z_k}{\ln r_e / r_w + s - 0.5} \right\} \tag{8.2}$$

for steady-state flow. For unsteady-state flow, 0.75 replaces 0.5. The factors α and λ_l are a unit conversion factor and a phase mobility, respectively and s is a skin factor. The pressure, p_k is that in grid block k at the drainage radius, r_e, where $r_e = \sqrt{\Delta x_i \Delta y_j / \pi}$.

We define the following:

---

† In this development, we ignore capillary-pressure effects in the well blocks; i.e., we assume $p_{ok} = p_{wk} = p_{gk} \equiv p_k$.

Special Concepts

$\bar{p}_{wf} \equiv$ The minimum bottomhole pressure constraint assigned to a datum point in the well, k_U.

$p_{wf} \equiv$ The actual flowing bottomhole pressure at k_U.

$\bar{q}_l \equiv$ A desired or specified rate for phase l ($l = o, w, g$). In some instances, the total liquid or fluid rates may alternatively be specified.

$q_l \equiv$ The actual well rate.

We consider two cases to illustrate the general approach; one where the well rates are specified; the other where the flowing bottomhole pressure is specified. It should be recognized that we are actually specifying boundary conditions in a multiply connected region R, i.e., each well within the reservoir R constitutes an interior boundary. When we assign a flowing bottomhole pressure to a well, we are specifying a Dirichlet condition there. Rate specifications are synonymous with Neumann boundary conditions. Both cannot be specified simultaneously over the same time period on a well. We can, however, switch from one to another in different time periods. For example, a rate or Neumann condition may exist for a certain period of time, after which it becomes a Dirichlet condition. This frequently occurs in pressure constrained wells where we desire production to occur at some fixed rate, provided the bottomhole pressure does not drop below a limiting value. When this is reached, the simulator will automatically maintain that pressure and the well produces the maximum rate possible with $p_{wf} = \bar{p}_{wf}$. Such a well is said to be *on deliverability*, i.e., it's delivering what it can while satisfying the pressure constraint. This is portrayed in Fig. 8.2.

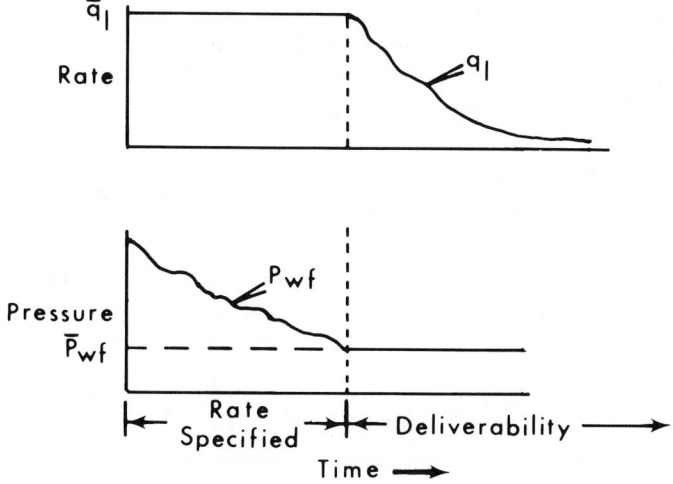

Fig. 8.2 Typical Well Specifications.

8.2.1 Production Rate Specified

The pressure p_{wfk} is related to that at k_U by (see Fig. 8.1.)

$$p_{wfk} = p_{wf} + \bar{\rho} \Delta d_k \tag{8.3}$$

where $\bar{\rho}$ is a mean fluid gradient (in pressure per unit distance) in the wellbore and $\Delta d_k \equiv d_k - d_{kU}$, a difference in elevations. Consequently, assuming the total oil rate is specified, the rate from layer k is

$$q_{ok} = PI_{ok}(p_k - p_{wf} - \bar{\rho}\Delta d_k) \tag{8.4}$$

where $PI_{ok} \equiv (WI \lambda_o)_k$, the productivity index. Adding the contributions from all the blocks in which the well is completed gives

$$q_o = \sum_j PI_{oj}(p_j - p_{wf} - \bar{\rho}\Delta d_j). \tag{8.5}$$

It follows that

$$p_{wf} = \frac{-\bar{q}_o + \sum_j PI_{oj}(p_j - \bar{\rho}\Delta d_j)}{\sum_j PI_{oj}} \tag{8.6}$$

where the sums in Eqs. 8.5 and 8.6 are from $j = k_U$ to $j = k_L$. This value of p_{wf} is the pressure at $k = k_U$ necessary to produce $\bar{q}_o$. If we find that $p_{wf} < \bar{p}_{wf}$, then the well cannot make the rate, $\bar{q}_o$, i.e., a lower pressure is required than is allowed. The well is then on deliverability, and the actual rate, $q_o < \bar{q}_o$. Thus, the production rate from layer k will be given by Eq. 8.4 where $p_{wf} = \max\{\bar{p}_{wf}, p_{wf}\}$ and p_{wf} is determined by Eq. 8.6.

If we specify an implicit rate on the well then we have

$$q_{ok}^{k+1} = q_{ok}^k + \frac{\partial q_{ok}}{\partial p_k}\delta p_k + \frac{\partial q_{lk}}{\partial S_{wk}}\delta S_{wk} + \frac{\partial q_{ok}}{\partial S_{gk}}\delta S_{gk} + \frac{\partial q_{ok}}{\partial p_{wf}}\delta p_{wf} \tag{8.7}$$

in terms of the four unknowns δp, δS_w, δS_g, and δp_{wf}. This is one more than we get from the flow equations (see chapter 7), consequently we need an additional relation. Since

$$\sum_j q_{oj}^{k+1} = \sum_j q_{oj}^k = \bar{q}_o,$$

we get the additional constraint

Special Concepts

$$\sum_j \left\{ \frac{\partial q_{oj}}{\partial p_j} \delta p_j + \frac{\partial q_{oj}}{\partial S_{wj}} \delta S_{wj} + \frac{\partial q_{oj}}{\partial S_{gj}} \delta S_{gj} + \frac{\partial q_{oj}}{\partial p_{wf}} \delta p_{wf} \right\} = 0. \tag{8.8}$$

Thus, when Eq. 8.8 is satisfied, we are guaranteed that the individual rates from each grid block in the vertical column between k_U and k_L add up to the total specified rate. If the value of p_{wf} as calculated by Eq. 8.6 is $\leq \bar{p}_{wf}$, then the nodal pressure at k_U is fixed which implies $\delta p_{wf} = 0$ and Eq. 8.8 is not required. When $p_{wf} > \bar{p}_{wf}$, we employ Eq. 8.8 to compute the change in flowing bottomhole pressure, i.e.,

$$\delta p_{wf} = \frac{\sum_j \left\{ \frac{\partial q_{oj}}{\partial p_j} \delta p_j + \frac{\partial q_{oj}}{\partial S_{wj}} \delta S_{wj} + \frac{\partial q_{oj}}{\partial S_{gj}} \delta S_{gj} \right\}}{\sum_j \frac{\partial q_{oj}}{\partial p_{wf}}}. \tag{8.9}$$

The partial derivatives in Eqs. 8.7–8.9 are determined by differentiating Eq. 8.4. Thus,

$$\frac{\partial q_{ok}}{\partial p_k} = PI_{ok}^k \tag{8.10}$$

$$\frac{\partial q_{ok}}{\partial S_{wk}} = \left\{ WI \left(\frac{b_o}{\mu_o} \right) \frac{dk_{ro}}{dS_w} (p - p_{wf}) \right\}_k^k \tag{8.11}$$

$$\frac{\partial q_{ok}}{\partial S_{gk}} = \left\{ WI \left(\frac{b_o}{\mu_o} \right) \frac{dk_{ro}}{dS_g} (p - p_{wf}) \right\}_k^k \tag{8.12}$$

$$\frac{\partial q_{ok}}{\partial p_{wf}} = -PI_{ok}^k \tag{8.13}$$

These derivatives are incorporated in the coefficient matrix **A** discussed in chapter 7. If lower than third degree implicitness is employed, then some or all of the derivatives in Eqs. 8.10–8.12 are zero. For example, for an implicit treatment of only p and S_w, the derivative in Eq. 8.12 is set to zero since gas saturation is treated explicitly (second degree implicitness). For an IMPES model, both Eqs. 8.11 and 8.12 are zero, except where semi-implicit rates are employed. Then one of these terms or its equivalent is retained during the saturation calculation.[1] If the rate is purely explicit then all the derivatives in Eq. 8.7 are zero.

A number of authors[2-5] employ schemes somewhat different from that outlined here. The major difference is that δp_{wf} is not carried as an additional unknown. When it is carried, there is considerable improvement in stability and time-step size capability.

If the well is not on deliverability, and the total fluid rate, $\bar{q}_t$, is specified, then we use a slightly different approach. Let f_{lk} be the fractional flow of phase l in block k defined by

$$f_{lK} = \frac{\lambda_{lk}}{\lambda_{ok} + \lambda_{wk} + \lambda_{gk}} \quad l = o, w, g. \tag{8.14}$$

Now $q_{lk} = f_{lk} q_{tk}$ and $\sum_j q_{tj} = \bar{q}_t$. We can write an expression similar to Eq. 8.7, substituting q_{lk} for q_{ok}. If we further assume that q_{tk} in each block is constant such that the constraint $\sum_j q_{tj} = \bar{q}_t$ is automatically satisfied, then we get

$$q_{lk}^{k+1} = q_{lk}^{k} + PI_{lk}\, \delta p_k + q_{tk}\left\{\frac{\partial f_{lk}}{\partial S_{wk}} \delta S_{wk} + \frac{\partial f_{lk}}{\partial S_{gk}} \delta S_{gk}\right\} + \frac{\partial q_{lk}}{\partial p_{wf}} \delta p_{wf}. \tag{8.15}$$

In Eq. 8.15, we let l take on its indices, o, w, and g and add the resulting three equations. This gives

$$q_{tk}^{k+1} = q_{tk}^{k} + (PI_{ok} + PI_{wk} + PI_{gk})\delta p_k + \frac{\partial q_{tk}}{\partial p_{wf}} \delta p_{wf}. \tag{8.16}$$

However, $\dfrac{\partial q_{tk}}{\partial p_{wf}} = -(PI_{ok} + PI_{wk} + PI_{gk})$. Consequently, the constraint condition leads to

$$\delta p_{wf} = \frac{\sum_j (PI_{oj} + PI_{wj} + PI_{gj})\delta p_j}{\sum_j (PI_{oj} + PI_{wj} + PI_{gj})}. \tag{8.17}$$

Then when the total rate is specified, Eq. 8.17 is employed to compute the change in flowing bottomhole pressure over a time step.

8.2.2 Specified Bottomhole Pressure

In the case when p_{wf} is specified, then $\delta p_{wf} = 0$ and Eqs. 8.9 and 8.17 are inactive. We need only compute q_{lk}^{k+1}, $l = o, w, g$. Consider flow of oil into the well for block k. Then the implicit rate calculation will employ Eq. 8.7 with the exception that the last term vanishes. To compute the associated water and gas, expressions similar to Eq. 8.7 are used. Alternatively, we proceed as follows: For water, we have

Special Concepts

$$q_{wk}^{k+1} = WI_k \left(\frac{b_w}{\mu_w}\right)_k^k \left(k_{rw}^k + \frac{dk_{rw}}{dS_w}\delta S_w\right)(p_k^{k+1} - p_{wfk}). \tag{8.18}$$

From Eq. 8.4, we have

$$\frac{q_{ok}^{k+1}}{PI_{ok}^k} = p_k^{k+1} - p_{wfk} \tag{8.19}$$

which when substituted into Eq. 8.18 gives

$$q_{wk}^{k+1} = \left(\frac{b_w\,\mu_o}{b_o\,\mu_w\,k_{ro}}\right)_k^k \left(k_{rw}^k + \frac{dk_{rw}}{dS_w}\delta S_w\right)q_{ok}^{k+1}. \tag{8.20}$$

An identical procedure for the gas phase yields

$$(q_{gf})_k^{k+1} = \left(\frac{b_g\,\mu_g}{b_o\,\mu_o\,k_{rg}}\right)_k^k \left(k_{rg}^k + \frac{dk_{rg}}{dS_g}\delta S_g\right)q_{ok}^{k+1} \tag{8.21}$$

where subscript f denotes free gas. The associated solution gas is given by

$$(q_{gs})_k^{k+1} = (q_o\,R_s)_k^k + q_o^n \frac{dR_s}{dp}\delta p \tag{8.22}$$

such that the total gas production becomes the sums of Eqs. 8.21 and 8.22.

8.2.3 Injection Wells

Our attention has been confined to production wells. Similar procedures are applied to injection wells. That is, for injection of water or gas, expressions like Eq. 8.7 are used to determine the allocation of the injected phase to block k, and the bottomhole pressure change is computed from expressions similar to Eq. 8.9. If injection is specified in a single-well block, as in a 2-D areal study, there is no allocation. One merely assigns the given rate as a source term to the injection block. When allocation does occur over a column of vertical blocks, a simulator should make it proportional to total mobility of all the flowing phases. The reason for this is that the injected phase must displace whatever phases are present in the injection block; i.e., the total mobility is the best measure of the total resistance.

If gas is injected, and the vertical communication is good, one can limit gas injection to the uppermost block. This is because gas rapidly percolates to the top of a column and overrides the oil. Such a practice can reduce computer costs substantially if there are a large number of gas injection wells.

8.3 Pseudo Functions

The expression "pseudo functions" was first applied in the reservoir simulation literature to describe modifications of laboratory-measured values of relative permeability and capillary pressure.[6] The modifications, in this case, were performed to reflect three-dimensional effects in a two-dimensional areal simulator. As happens so often when a new concept is introduced, the name adopted to describe it conveys the wrong impression. The same is true in this instance. The expression, "pseudo functions" perhaps more aptly describes laboratory measured rock and rock-fluid parameters, including those cited above. What are called pseudo functions are probably closer to the true reservoir parameters represented by their laboratory surrogates. Nevertheless, for purposes of clarity and common understanding, we speak of pseudo functions as rock and/or rock-fluid properties that have been modified from their laboratory counterparts to achieve some particular goal. Frequently, pseudo functions are identified with pseudo relative permeabilities and pseudo capillary pressures only. Here we adopt a slightly broader outlook and apply the expression to any modified rock and fluid property including those above. Thus we can have pseudo permeabilities, porosities, saturations, potentials, rates, viscosities, etc. These pseudo functions are consistently identified here by a superior tilde, e.g., $\tilde{\Phi}$, $\tilde{S}_w$, etc.

In most instances, pseudo functions are weighted averages of a parameter over a reservoir-size volume. By reservoir-size volume we mean an elemental volume of the reservoir, but on a larger scale than that used to make laboratory measurements of rock properties. For example, a collection of grid blocks, or an individual grid block, in a reservoir simulator is a reservoir-size volume. We will also have occasion to speak of "rock relative permeabilities," "rock capillary pressures," or for short, "rock parameters." These terms are used with reference to all rock, fluid, and rock-fluid parameters measured in the laboratory. This, of course, involves a much smaller volume than a reservoir-size volume. One can appreciate that permeability, for example, measured on a small core sample differs significantly from the average value over the reservoir-size volume from which the sample was taken. The first is simply "permeability," the second is "pseudo permeability" if the terminology of pseudo functions is applied consistently.

Historically, pseudo functions came about by considering the following question: Given a reservoir which is recognizably a three-dimensional entity, how can we simulate its behavior with a two-dimensional areal model? For such a model, the reservoir-size volumes for which we derive weighted average properties, or pseudo functions, are characterized by one grid-block in the dip-normal direction. If the pseudos are to reflect the conditions that exist in the actual reservoir, they must somehow convey effects in the neglected dip-normal (or vertical) direction. This usually necessitates various assumptions about the distribution and state of fluids along this coordinate axis. Pseudo functions derived for reservoirs satisfying these conditions, therefore, embody or "capture" those effects in the excluded third dimension, and transfer them to the areal simulation plane. Furthermore, they act as transformations that permit reduction of the dimensionality of a reservoir problem.

Recent developments have made it possible to endow pseudo functions with characteristics other than those cited above. For example, if the reservoir-size volumes, or grid block sizes used by a simulator, are made small when the pseudo functions are generated, then they will transfer the effects of this grid refinement even when they are subsequently used in a coarse grid model.[7-9] That is, the better accuracy associated with the fine grid is captured by the pseudos and conveyed to the coarse grid. Furthermore, pseudo functions generated by one simulator and used in a completely different simulator reflect the characteristics of the parent-model.

In summary, we can think of pseudo functions in simplest terms—they are weighted averages of certain reservoir engineering variables; they are transformations insofar as they reduce dimensionality of simulation problems; they have a gene-like quality to the extent they transfer to the model that uses them, certain traits of the model that produced them.

8.3.1 Pseudo Functions Based on the Vertical Equilibrium Concept[6,10]

Perhaps the best way to understand the VE concept is to visualize a horizontal reservoir containing oil, water, and gas with distinct contacts between phases. When we speak of VE, we refer to a condition of fluid distribution in this system that is assumed to persist, or essentially so, over its productive life. Most reservoirs at discovery, and prior to being produced, are in an equilibrium state. By this we mean that a traverse vertically downward would encounter a balance of forces at all points resulting in a static equilibrium condition. Even after production ensues, vertical disturbances from this initial condition may be slight except for a more or less uniform migration of fluid contacts. When these disturbances occur, the existence of a VE condition means they are rapidly dissipated relative to the movement toward wells in the horizontal plane. Obviously, this will occur only if there is good vertical communication. Actually, vertical equilibrium is not the same as an assumption of no-flow in the vertical direction. Rather, it is the equivalent of an assumption of infinite vertical flow rate, such that the time required for vertical transients to decay is zero. Under such conditions, when an equilibrium state is reached, the vertical potential gradients in a fluid phase become zero. Thus, the pressure will vary hydrostatically, in a phase, along any vertical coordinate direction.

We categorize various types of VE based on the forces acting on the fluids, that is, the viscous, capillary and gravity forces. The major effects of the capillary and gravity forces are usually associated with the vertical direction, while the principal viscous force is frequently lateral, except locally around wells. A VE condition is approximated when the vertical viscous force is small, and there is a balance of gravity and capillary forces. We refer to this situation as *gravity-capillary vertical equilibrium*. Now, if the capillary forces should be insignificant relative to the gravity forces, the fluids will migrate to another equilibrium state—one where the fluids are segregated. We refer to this as *gravity-segregated vertical equilibrium*. These two VE situations are the only ones we need consider. Another type of VE can exist where the lateral viscous forces dominate the capillary, gravity, and vertical viscous forces in a stratified system.[11] However, it can be shown that this is nothing more than a special case of gravity-segregated VE.[12]

Conditions of vertical equilibrium are approximated in reservoirs having one or more of the following properties:

(1) High vertical permeabilities
(2) High gravity and/or capillary forces
(3) High fluid mobilities
(4) Low rates of areal fluid movement
(5) Low vertical potential gradients

In the following sections, the VE concept is treated in more detail.

8.3.2 Gravity-Capillary Vertical Equilibrium

We first consider an equilibrium condition as a result of a gravity-capillary balance. For this purpose, we treat a reservoir having an average bed thickness, h, containing water and oil with water as the wetting phase. Throughout the entire thickness, the fluids are in vertical equilibrium with the typical capillary pressure curve depicted in Fig. 8.3.

Initially, we consider the capillary transition zone is an appreciable part of the total thickness. The mathematical conditions for vertical equilibrium are[6]

$$\frac{\partial \Phi_w}{\partial z} = \frac{\partial \Phi_o}{\partial z} = 0 \qquad (8.23)$$

$$P_c(x,y,z) = \tilde{P}_c(x,y) - \cos\theta \int_0^z \Delta\gamma\, dz \qquad (8.24)$$

where z is the coordinate direction that is normal to the bedding plane, i.e., along the dip-normal line, θ is the dip angle, and $\delta\gamma \equiv \gamma_w - \gamma_o$, the

Special Concepts

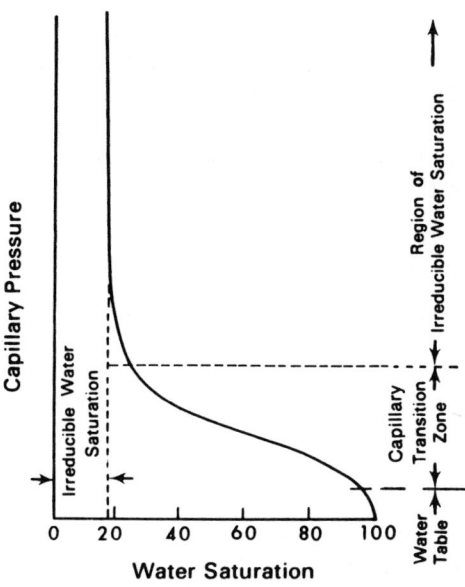

Fig. 8.3 Typical Water-Oil Capillary Pressure Curve for Gravity-Capillary VE.

specific density difference. In Eq. 8.23, $\tilde{P}_c\,(x,y)$ is a capillary pressure taken at any areal point (x,y) on a reference surface passing through the reservoir and $P_c\,(x,y,z)$ is the capillary pressure at some point in the 3-D medium. For convenience, we let the reference surface coincide with the top of the reservoir as shown in Fig. 8.4.

The flow of water and oil at a point, P, in this system is described by

$$\frac{\partial}{\partial x}\left\{\frac{k_{xy}k_{rl}}{\mu_l}\frac{\partial \Phi_l}{\partial x}\right\} + \frac{\partial}{\partial y}\left\{\frac{k_{xy}k_{rl}}{\mu_l}\frac{\partial \Phi_l}{\partial y}\right\} = \frac{\partial}{\partial t}(\phi S_l) \qquad (8.25)$$

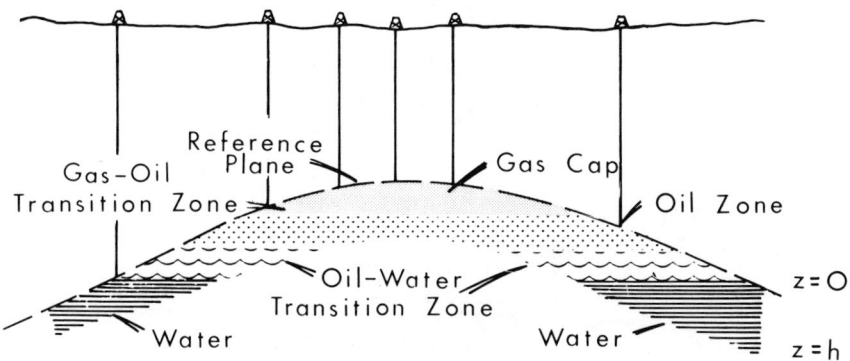

Fig. 8.4 Typical Reservoir Showing Reference Plane Location.

where $\mathit{l} = w, o$ (water or oil) and we assume no sources or sinks (wells) occur at P. Here k_{xy} is the absolute permeability in the x-y plane. In general, k_{xy} and the saturation, S_l, under a VE condition will vary with the dip-normal coordinate, z. This S_l-dependency means the relative permeability, k_{rl}, is an implicit function of z. Thus, integrating Eq. 8.25 with respect to z over the reservoir thickness gives

$$(\Phi_{lx} I_1 / \mu_l)_x + (\Phi_{ly} I_1 / \mu_l)_y = (I_2)_t \tag{8.26}$$

where partial differentiation is denoted by the subscripts x, y, and t, i.e.,

$$\Phi_{ls} \equiv \frac{\partial \Phi_l}{\partial s} \text{ and } (\cdot)_s \equiv \frac{\partial}{\partial s}(\cdot), \, s = x, \, y \text{ or } t.$$

Also, we have

$$\left. \begin{aligned} I_1 &\equiv \int_0^h k_{xy}(z) \, k_{rl}(S_l) \, dz \\ I_2 &\equiv \int_0^h \phi(z) \, S_l(z) \, dz \end{aligned} \right\} . \tag{8.27}$$

To arrive at Eq. 8.26, it is necessary to assume the viscosity, μ_l, is invariant with depth. Also, orders of differentiation and integration on the right-hand side have been interchanged.

The integrals in Eq. 8.27 are evaluated by the first mean value theorem (see Appendix A.3) yielding $I_1 = \tilde{k}_{xy} \tilde{k}_{rl} h$ and $I_2 = \tilde{\phi} \tilde{S}_l h$ where the quantities with the tildes are average values in the interval $(0, h)$. In fact,

$$\left. \begin{aligned} \tilde{k}_{xy} &= \frac{1}{h} \int_0^h k_{xy}(z) \, dz \\ \tilde{\phi} &= \frac{1}{h} \int_0^h \phi(x) \, dz \end{aligned} \right\} . \tag{8.28}$$

Consequently,

$$\tilde{k}_{rl} = \frac{\int_0^h k_{xy}(z) \, k_{rl}(z) \, dz}{\int_0^h k_{xy}(z) \, dz}, \, \mathit{l} = w, o \tag{8.29}$$

$$\tilde{S}_l = \frac{\int_0^h \phi(z) S_l(z)\, dz}{\int_0^h \phi(z)\, dz}, \quad l = w, o \qquad (8.30)$$

Eqs. 8.23–8.24 and Eqs. 8.29–8.30 are the VE equations originally given by Coats et al.[6] As these authors show, Eq. 8.23 can be differentiated with respect to z yielding $dz = dP_c/\Delta\gamma \cos\theta$. When this is substituted in the numerators of Eqs. 8.29 and 8.30 the result is

$$\tilde{k}_{rl} = a \int_{\tilde{P}_c}^{\tilde{P}_c + \Delta P_c} k_{xy} k_{rl}\, dP_c \Big/ \int_0^h k_{xy}\, dz \qquad (8.31)$$

$$\tilde{S}_l = a \int_{\tilde{P}_c}^{\tilde{P}_c + \Delta P_c} \phi S_l\, dP_c \Big/ \int_0^h \phi\, dz \qquad (8.32)$$

where $a = 1/(\Delta\gamma \cos\theta)$ and $\Delta P_c = h\Delta\gamma \cos\theta$. These equations are used to arrive at the pseudo relative permeabilities, $\tilde{k}_{rl}$, and pseudo capillary pressures, $\tilde{P}_c$, as functions of the volumetrically averaged saturation (or pseudo saturation), $\tilde{S}_w$. Thus, for a given value of z where $0 \leq z \leq h$, $\tilde{P}_c$ is calculated from Eq. 8.23 using the rock-curve value of P_c corresponding to S_w at z. Then $\tilde{k}_{rl}$ and $\tilde{S}_w$ can be computed readily from Eqs. 8.31 and

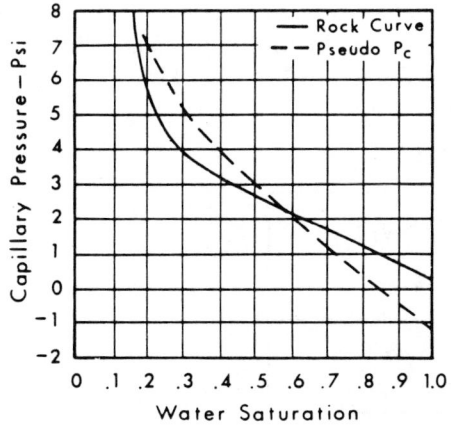
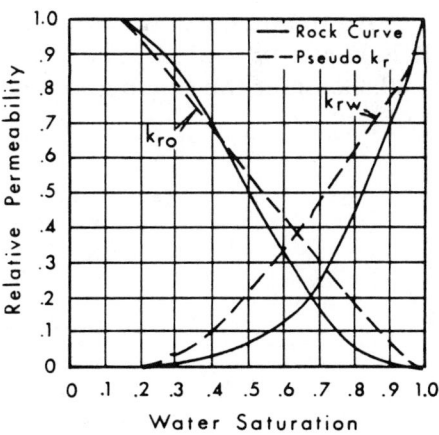

Fig. 8.5 Typical Gravity-Capillary VE Pseudo Functions (after Coats et al.[6]).

8.32 since they are single-valued functions of P_c. Once $\tilde{S}_w$ is obtained from Eq. 8.32, then $\tilde{S}_o = 1 - \tilde{S}_w$.

Typical pseudo relative permeability and capillary pressure curves are shown in Fig. 8.5 for a gravity-capillary VE condition. Note they tend to be more linear than the original rock curves. This tendency increases as h increases. In fact, if h increases to the point where the capillary transition zone (Fig. 8.3) is negligibly small, then the pseudocurves become straight line segments for fully segregated fluids.

These pseudo functions transfer the effects of the dip-normal direction into the x-y simulation plane. Consequently, one reaps the advantage of a less expensive 2-D computation while accounting for the third dimension. Another advantage is improved convergence rates when an iterative procedure is used by the simulator. This is because convergence is directly linked to the nonlinearities introduced by the rock P_c and k_r-curves. The linearization gained by using the pseudo functions means convergence occurs in fewer iterations. Thus, the overall effect is to reduce the computational requirements substantially by as much as 70 percent in some cases.[6]

8.3.3 Gravity-Segregated Vertical Equilibrium

If the capillary transition zone is less than 10 percent of the total reservoir thickness, then gravity-segregated VE occurs. Such a condition was first treated by van Poollen, Breitenbach and Thurnau.[13] They compute an average or pseudo relative permeability that can be used to approximate the third dimension based on material balance considerations. A similar approach was used to treat cases where there was partial dispersion, i.e., where a previously segregated zone disperses into another. A more comprehensive treatment of gravity-segregated VE was provided by Coats et al.[10] It is based on a piece-wise integration of the formulas in Eqs. 8.29 and 8.30. Consider the single grid block represented in Fig. 8.6. Here z_{ci} repre-

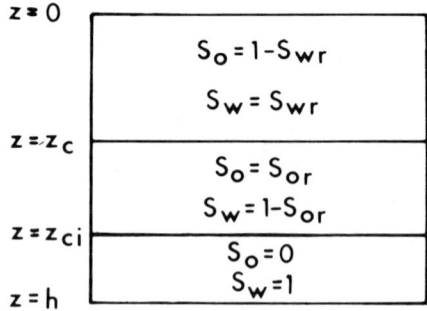

Fig. 8.6 Segregated Fluids in a Single Grid Block.

Special Concepts

sents the original water-oil contact and z_c is its location at a later time. For the condition shown here:

$$\tilde{S}_w = \{z_c S_{wr} + (1 - S_{or})(z_{ci} - z_c) + (h - z_{ci})\}/h \qquad (8.33)$$

$$\tilde{k}_{rw} = \{k_{rw}^*(z_{ci} - z_c) + h - z_{ci}\}/h \qquad (8.34)$$

$$\tilde{k}_{ro} = k_{ro}^* z_c/h \qquad (8.35)$$

where $k_{rw}^* = k_{rw}(1 - S_{or})$ and $k_{ro}^* = k_{ro}(S_{wr})$ for irreducible water and oil saturations of S_{wr} and S_{or}, respectively. We refer to these values of relative permeability as the *end-point values* (see Fig. 3.3).

Eq. 8.24 is used to define a pseudo capillary pressure function where we assume the rock's capillary forces are zero. If the specific density difference is constant within a cell and $\alpha = 0$, then

$$\tilde{P}_c = \Delta \gamma z_c \qquad (8.36)$$

where the upper limit on the integral is taken as z_c in Eq. 8.24 since its changes with time define the various values of $\tilde{S}_w$. Hence, one can construct $\tilde{P}_c$ as a function of $\tilde{S}_w$. Observe that here $\tilde{P}_c$ has no relationship to actual capillary forces. In this case, the name "pseudo capillary pressure" is justified.

These pseudo functions, i.e., the pseudo capillary pressure and pseudo relative permeability appear as straight line segments in terms of the pseudo saturation $\tilde{S}_w$. When the reservoir has variable structure, then these formulas still apply, except that we must make some different substitutions for the variables h and z. In general, in this case each grid block in the simulator will have its own unique set of pseudo functions.[10] If the reservoir is stratified with distinct values of porosity, permeability and saturation, which are constant in each layer, then the integrals in Eqs. 8.29 and 8.30 become discrete sums. Coats et al.[10] provide formulas for a three-layer case. These can readily be generalized for an n-layer situation.[12] From such a generalization, it is a simple matter to derive the formulas of Hearn.[11,12]

8.3.4 Validation of Vertical Equilibrium

When does a reservoir reasonably approximate VE conditions? Certain criteria have been proposed based on various mathematical procedures.[6,10] For the most part, these criteria are not reliable. Validity is generally determined by first running cross-sectional studies; pseudo functions are then generated from these results. The simulator is run again in a 1-D areal mode using the pseudo functions. If the results compare favorably with the cross-sectional runs, one knows the pseudo functions adequately reflect

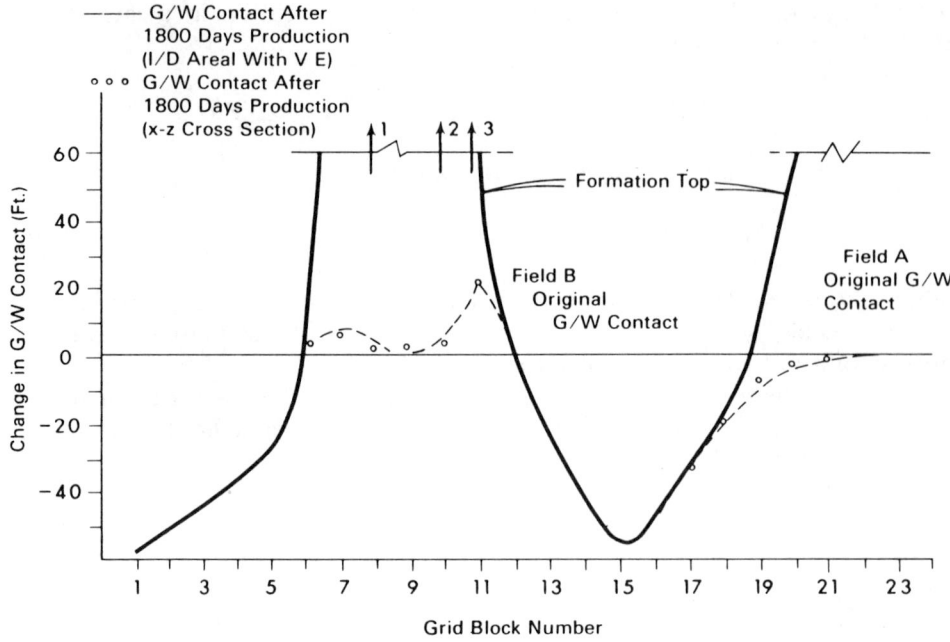

Fig. 8.7 Comparison of 2-D Cross Section and 1-D VE Results (After Coats et al.[10]).

the z-direction. The pseudo functions are then used in the model in a 2-D areal mode to simulate 3-D effects.

An example is shown in Fig. 8.7 (After Coats et al.[10]). In this case, Field B was producing gas through three wells, while Field A was shut in. The substantial closure and areal extent of the reservoirs normally would require 3-D treatment. However, because of the density contrasts between the fluid phases, one can surmise that a segregated VE condition exists and the problem can be handled with a cheaper 2-D simulation. Fig. 8.7 demonstrates the validity of this supposition. The computing times for the 1-D and 2-D simulations were 3 seconds and 74 seconds, respectively, for the 1800-day run. Projecting these to a 3-D study, it is estimated that 25 times more machine time would be required than a 2-D areal study using VE.

8.3.5 Dynamic Pseudo Functions

When a reservoir does not maintain a vertical equilibrium condition over most of its life, then the concepts discussed thus far are unreliable. Substantial deviations from a VE condition occur when there are large

Special Concepts

changes in flow rates as the field is developed. If vertical communication is poor, the vertical disturbances will not be rapidly dissipated relative to lateral fluid movement. Thus, appreciable dip-normal potential gradients can exist, in violation of a basic premise of VE. Such a reservoir is characterized by fluid distributions that undergo large changes with time. Unlike VE, which presupposes an essentially static vertical distribution, the reservoir is in a dynamic condition. Obviously, if pseudo functions are to be employed in a 2-D areal simulation, they must reflect the dynamic condition of the reservoir. This means they must embody the time-dependent history of the vertical coordinate. Consequently, we can view dynamic pseudo functions as entities that retain a "memory" of the history of a coordinate direction.

One approximation to a dynamic situation in a reservoir is to treat it as a sequence of VE states, each assumed to persist for a short period of time. To arrive at these states, it is necessary to perform a number of representative 2-D cross-sectional simulations that span the rate conditions expected in the 2-D areal simulation. At any moment in time, a printout of the vertical saturation distribution for a column of blocks in the cross-sectional model can represent a VE condition. In general, it will differ from that at another moment in time or another column of blocks. For each state, one can apply Eqs. 8.24, 8.29, and 8.30 (or their specialized forms) to arrive at sets of pseudo functions which are specific for a given time. In practice, differences between columns are frequently low so that the pseudos can be lumped or averaged for a collection of columns.

Once the different sets of pseudo functions are generated, they are correlated to tie the different equilibrium states together. This is necessary before the pseudos can be used. Such a procedure is essentially one of providing continuity to the pseudo functions in the time domain. Logical parameters to use in such correlations are fluid velocities since they largely determine the differences between the VE states for fixed reservoir conditions and well distributions. Once such correlations are developed, they are used to determine the appropriate pseudos as functions of time in a 2-D areal calculation. Prior to this, however, validation runs should be performed in 1-D with the dynamic pseudos. The results should compare favourably with those from the cross-sectional runs. Jacks et al.[14] originated this approach and successfully applied it to a large carbonate reservoir. They also demonstrated the validity of their dynamic pseudos by using them in both 1-D and 2-D areal models and compared the results with 2-D cross-sectional and 3-D simulations, respectively. This technique has also been successfully exploited by Stright.[15]

Kyte and Berry[7] present a different approach for computing dynamic pseudos. Implicit in their procedure, though not stated, is the assumption expressed by Eq. 8.23. By considering total resistance to flow through several rows of stacked grid cells, and relating this to an Ohm's Law analogy of

Darcy's Law, a type of harmonic average† permeability is computed for the collection of cells. Within this collection, pseudo pressures $\bar{p}_l$ are also computed, corrected to the centers of the group. Pseudo capillary pressures are determined by taking differences in the pseudo phase pressures. Pseudo relative permeabilities are computed by Darcy's Law, expressed in terms of the pseudo pressures and the approximation to the harmonic average permeability. The nice thing about this approach is that the pseudo functions capture the grid refinement of the model that generates them. Thus, if they are generated in a refined grid cross-sectional model (the parent model) and are subsequently employed in a coarse grid areal model, they convey to the latter the accuracy inherent in the parent model. Thus, they can be used to control *numerical dispersion effects*;[7] i.e., they minimize the smearing of saturation and concentration fronts that results from discretization of the pertinent partial differential equations. This aspect of pseudo functions has been exploited by Breit and Graue[9] and Camy and Emanuel[8] in simulating enhanced oil recovery processes.

To illustrate the gene-like quality of the Kyte-Berry pseudo functions, we reproduce some of their results in Fig. 8.8. The problem was a 27 x 20 cross section (depicted in the top of the figure), with an oil producer in block 1 and a water producer in block 27. This cross section constituted the parent model for generating pseudo relative permeabilities and capillary pressures. Thereafter they were used in a 9 x 1 areal analog of the cross section (middle of the figure). The results are shown in the bottom of the figure where the saturation distribution is plotted at breakthrough. Also shown on this plot are the dynamic pseudo functions of Jacks et al.[14] assuming a sequence of VE states (labeled as "conventional pseudos"). Notice the Kyte-Berry pseudos (labeled as "new") capture the grid-refinement, whereas the Jacks et al. pseudos do not. Furthermore, the 9 x 1 system yields recovery at breakthrough equal to that of the cross section even though there is a 60-fold reduction in the number of grid blocks employed.

8.3.6 Well Pseudo Functions

Because of converging flow patterns and coning phenomena associated with wells, special techniques are required if one desires to capture these effects in well blocks of a cartesian simulator. There are three methods for doing this:[16-18]

(1) Project the effect of vertical flow by way of pseudo relative permeabilities into the well blocks of a 2-D areal simulator.[16]

---

† It can be shown this is not a true harmonic average but rather yields a value less than the harmonic average.[12]

Special Concepts

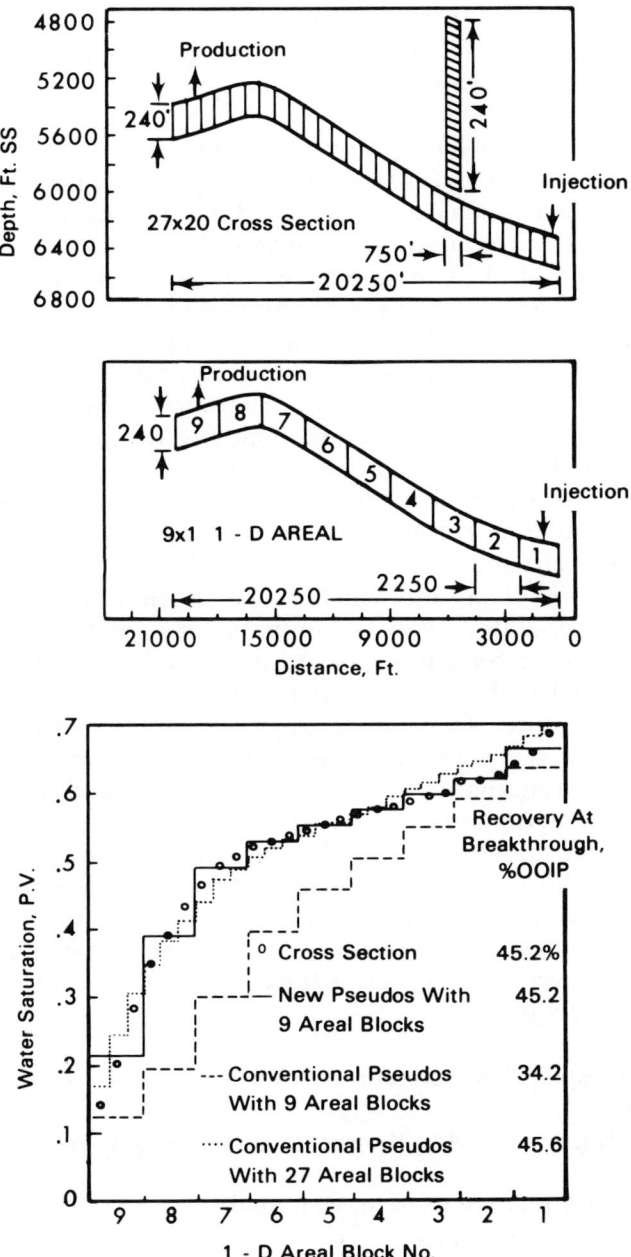

Fig. 8.8 Comparison of Results from a 2-D and 1-D Simulation (After Kyte and Berry[7]).

(2) A semi-analytical procedure.[17]
(3) Transplantation of pseudo functions generated by a well coning model into a cartesian model.[18]

The first method essentially captures the vertical flow performance around a well. The final result is a pseudo relative permeability curve which is used in the well blocks of a cartesian simulator.[16] This curve is usually different from the rock curves used in the interwell regions.

The semi-analytical technique is actually an extension of an idea first proposed by Fagin et al.[19] The pseudo function in this case takes the form of an analytical expression. The main disadvantage of this treatment is that it is extremely difficult to obtain general analytical expressions to capture the well effects we would like to include. Therefore, simplifying assumptions are required. This renders the method highly restrictive and not generally applicable to a wide variety of situations. Chappelear and Hirasaki[17] applied this technique to a water-oil system in which they wanted to include water coning effects. We don't recommend this approach because it is not readily adapted to other circumstances.

Our treatment of pseudo functions has been directed toward cases where they are generated by the same simulator in which they are eventually used. One can extend this concept and employ a different model to construct them with the intent of more accurately depicting the effects one wishes to capture. The pseudo functions are then transplanted into the simulator that will ultimately use them. A single-well 2-D cylindrical model is capable of simulating typical converging flow and coning characteristics. If pseudo functions are generated in such a model, then they transfer these effects when used in the well blocks of a cartesian model.

Woods and Khurana[18] introduced this concept for cases involving water coning. They obtained mathematical expressions for pseudo relative permeability and pseudo capillary pressure by invoking similarity principles between a coning model and a 3-D cartesian reservoir model. The following assumptions are required:

(1) In corresponding volumes in each model, the net mass fluxes are equal.
(2) In each model having the same volume, the average potentials are identical.
(3) Corresponding volumes in each model have the same average fluid saturations at any time.

With these similar relationships one computes pseudo relative permeabilities and capillary pressures as functions of a pore volume-weighted saturation. These induce within a well column of rectangular blocks in a cartesian model the same vertical saturation and potential distributions observed in

Special Concepts

the cylindrical coning model. Results of such an application are given by Woods and Khurana[18]; however, the idea has not been fully exploited.

8.4 Gas Percolation

In models offering first degree implicitness (IMPES models), an unstable situation arises when gas comes out of solution and travels upstructure. Because gas viscosity is low, the mobility is high, and the volume of gas transferred over a reasonable time-step size can be many times the pore volume of any block it passes through. This leads to negative saturations unless severe restrictions are placed on the time-step size. The instability is due to the explicit treatment of the transmissibilities. With the use of semi-implicit transmissibilities in an IMPES model, or models offering higher degrees of implicitness we don't encounter the problem. For IMPES models with explicit transmissibilities, schemes have been devised that permit one to avoid gas percolation instabilities.[20,21]

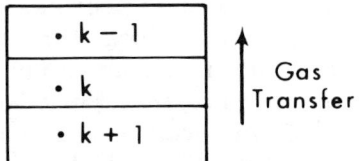

Fig. 8.9 Gas Percolation.

To illustrate the problem, consider the 1-D vertical system of blocks given in Fig. 8.9. We assume $\Phi_{k+1} > \Phi_k$ where here Φ is the gas-phase potential. The quantity of gas flowing from $k+1$ to k during a time-step Δt is

$$Q_g = (T_g)_{k+1/2} \left(\frac{k_{rg} b_g}{\mu_g} \right)_{k+1/2} (\Phi_{k+1}^{n+1} - \Phi_k^n) \Delta t. \qquad (8.37)$$

The mobile gas content at the beginning of the time step in the $(k+1)^{st}$ cell is

$$GIP = \{ V_p (S_g - S_{gr}) b_g \}_{k+1}^n. \qquad (8.38)$$

If $Q_g > GIP$, then S_g will become negative in the lower block and the instability becomes worse on succeeding time steps. To prevent this from happening one can invoke the constraint, $Q_g / GIP < 1$, which implies

$$\Delta t \leq \frac{V_p \, (S_g b_g)^n}{(T_g)_{k+1/2} \, \lambda_g^n (\Delta \Phi^{n+1})}, \qquad (8.39)$$

$\lambda_g \equiv k_{rg} \, b_g / \mu_g$, $\Delta \Phi \equiv \Phi_{k+1}^n - \Phi_k^{n+1}$. This relation then becomes a stability condition, i.e., to prevent instability, the inequality in Eq. 8.39 must be preserved. Usually, this constitutes a severe restriction on the time-step size. Even if Δt satisfies Eq. 8.39, the quantity Q_g will still be in error since S_g is rapidly decreasing in block $k + 1$.

A number of commercial IMPES models permit the user to invoke a special gas percolation routine when rapid upstructure gas transfers occur. Some employ a multiplicative factor, β say, where $0 < \beta < 1$ on the interface gas transmissibilities to retard gas flow. The difficulty with this approach is that there is no physical rationale one can employ to determine the value of β. Typically, values between 0.1 and 0.2 are used. Another approach is to compute a factor $K_v = Q_g / GIP$ at each time step. If $K_v > 1$, the value of T_g is modified such that $K_v = 1$, i.e., the maximum flow of gas is limited to the mobile content of the block from which it comes. A third approach is also based on monitoring K_v. When it exceeds one, T_g between adjacent grid blocks is set to zero and a source term in the amount of the gas content in the lower block is introduced to the upper block in a column. A similar sink term is introduced in the lower block. The reason for this treatment is that gas percolation culminates in accumulation of the total transferred volume upstructure over an extremely short period of time relative to typical time step sizes (30–90 days) employed in a simulator. Thus, rather than perform the transfer by a mathematical calculation that leads to instability, a physical transfer is used by way of source/sink terms.

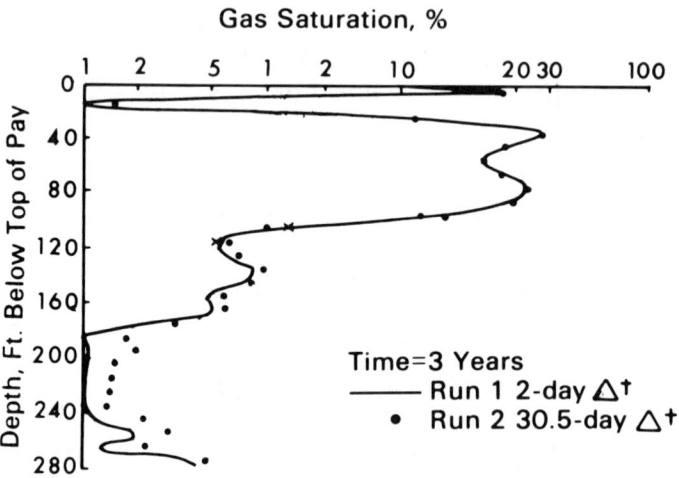

Fig. 8.10 Depth Variation of Gas Saturation from a 1-D Simulation (After Coats[20]).

Special Concepts

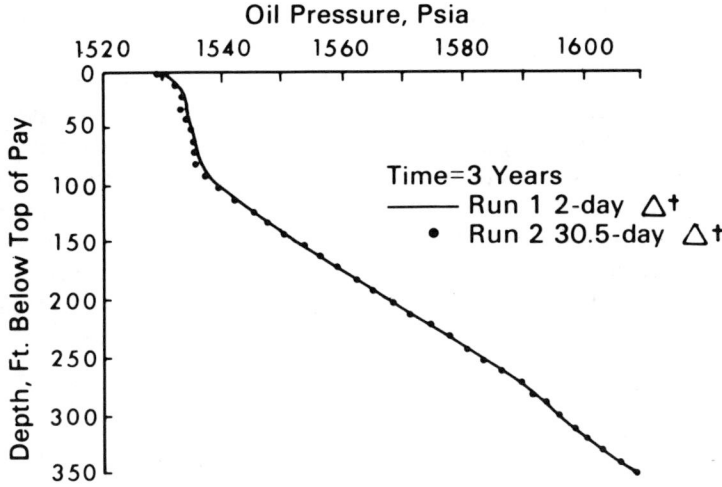

Fig. 8.11 Depth Variation of Pressure from a 1-D Simulation (After Coats[20]).

These techniques usually work satisfactorily. For example, some results are shown in Figs. 8.10 and 8.11. The application was to a pinnacle reef in Alberta which was modeled in 1-D (vertically). Fig. 8.10 shows the calculated gas saturation vs. depth while Fig. 8.11 displays the oil saturation. The solid lines are for a time-step size of 2 days. In this case, nearly all blocks flowed less gas than they held. The dotted line is for $\Delta t = 30.5$ days where in 26 of the total 36 vertically stacked grid blocks, a special gas percolation routine was used.[20] The agreement between the two cases is quite good.

8.5 Variable Bubble-Point Problems

Recall in chapter 3 that the saturation pressure, p_s, is defined as the minimum pressure required to dissolve a given volume of gas in oil. A slight reduction in pressure below p_s evolves the first bubble of gas, hence it is also called the *bubble-point pressure*. If gas is injected in an undersaturated reservoir it will dissolve, causing a shift to a higher value of p_s. A shift to a value lower than the original occurs, if after a period of production in which gas is evolved and produced or accumulated upstructure, the reservoir is repressured by water injection. The lower value is in the downstructure region, while a higher value can occur above. Moreover, reservoirs of substantial thickness are sometimes characterized by large variations of p_s with depth. In the developments in chapter 7 there is a tacit assumption that p_s remains fixed. This is usually adequate if, during depletion, pressure decreases everywhere in the reservoir. However, for the situations cited above variability must be taken into account.

Some methods for treating variable bubble-point pressures involve adjusting the accumulation terms, i.e., the right-hand sides of Eq. 7.64 or Eq. 7.108.[22-24] A fully implicit treatment presented by Stright et al.[25] is the most rigorous approach and is incorporated in a natural way in an implicit simulator such as that discussed in section 7.3.3. The technique relies upon distinguishing between bubble-point pressure and saturation pressure. Saturation pressure is identical with the grid block pressure if there is free gas in the block. If, on the other hand, the cell is undersaturated, then p_s is identified as the bubble-point pressure. The following conditions then apply:[25]

(1) $p \geq p_s$
(2) If $p > p_s$, $S_g = 0$
(3) If $S_g > 0$, then $p = p_s$

Consequently, p_s and S_g are mutually exclusive variables which means variable substitution can be employed. For example, when $S_g^k = 0$, the computation of δS_g is redundant since $\delta S_g = S_g^k$, a known value at the beginning of an iteration. When this occurs, δp_s is substituted for δS_g as an implicit variable.

To accomplish the computation of δp_s it is necessary to express b_o and R_s in terms of p_s. This is done using $b_o = b_o(p, p_s)$ and $R_s = R_s(p_s)$. The consequence of this is that all terms involving b_o in Eqs. 7.125 and 7.133 are expanded accounting for the dependence of b_o on p and p_s. Those involving R_s remain the same as given in chapter 7 with the exception that R_s is interpreted as a derivative with respect to p_s rather than p (for example, Eq. 7.121). Stright et al.[25] give an algorithm for expanding the right-hand sides considering three possible situations. A simpler approach is given by Thurnau and Thomas for implicit simulators.[26]

8.6 Exercises

1. (a) Show that Eqs. 8.33–8.35 follow from Eqs. 8.29–8.30.
   (b) Establish the corresponding formulas if $z_c > z_{ci}$, i.e., if the water-oil contact moves down.
2. Derive an explicit formula showing the dependence of $\tilde{P}_c$ on $\tilde{S}_w$ for gravity-segregated VE when $z_c < z_{ci}$ and when $z_c > z_{ci}$.
3. How would the relationships obtained in problem 2 appear if you were to plot $\tilde{P}_c$ as a function of $\tilde{S}_w$ for $0 \leq \tilde{S}_w \leq 1$? Provide a sketch with your answer.

4. Suppose a shallow, stratified reservoir, consisting of five layers is in gravity-segregated VE. Pertinent data are given in the table below.

Reservoir Properties

Layer No.	Permeability k, md.	Porosity ϕ	Thickness h, ft.	Depth z, ft.
1	27	0.18	30	30
2	45	0.15	55	85
3	37	0.23	61	146
4	114	0.09	109	255
5	56	0.12	65	320

The initial water-oil contact is located at $z_{ci} = 230$ feet. End-point data from the relative permeability curves are $S_{wr} = 0.23$, $S_{or} = 0.30$, $k_{ro}^* = 0.83$, and $k_{rw}^* = 0.41$. Let z_c successively take on values of 280, 230, 186, 115, and 75 feet.
  (a) Beginning with Eqs. 8.29 and 8.30 determine the discrete equations required to compute $\tilde{k}_{rw}$, $\tilde{k}_{ro}$, and $\tilde{S}_w$.
  (b) Using the data above, compute $\tilde{k}_{rw}$ and $\tilde{k}_{ro}$ as functions of $\tilde{S}_w$ and plot. What peculiarities do you observe? What are the reasons for these phenomena?
  (c) If $\Delta\gamma = 0.086$ psi/ft. plot $\tilde{P}_c$ as a function of $\tilde{S}_w$.
5. Derive a general form for the C-coefficients in chapter 7 assuming a variable bubble-point pressure. An exact algorithm is not required.

8.7 References

1. MacDonald, K.C. and Coats, K.H.: "Methods for Numerical Simulation of Water and Gas Coning," *Trans.*, AIME (1970) **249**, 245.
2. Chappelear, J.E. and Rogers, W.: "Some Practical Considerations in the Construction of a Semi-Implicit Simulator," *Soc. Pet. Eng. J.* (June 1974) 216.
3. Chappelear, J.E. and Williamson, A.S.: "Representing Wells in Numerical Reservoir Simulation—Theory and Implementation," paper SPE 7697 presented at the 5th Symposium on Reservoir Simulation, Denver, Jan. 31–Feb. 2, 1979.
4. Nolen, J.S. and Berry, D.W.: "Tests of the Stability and Time-Step Sensitivity of Semi-Implicit Reservoir Simulation Techniques," *Trans.*, AIME (1972) **253**, 253.
5. Aziz, K. and Settari, A.: *Petroleum Reservoir Simulation*, Applied Science Publishers, London (1979).
6. Coats, K.H., Nielson, R.L., Terhune, M.H., and Weber, A.G.: "Simulation of Three-Dimensional, Two-Phase Flow in Oil and Gas Reservoirs," *Soc. Pet. Eng. J.* (Dec. 1967) 377.
7. Kyte, J.R. and Berry, D.W.: "New Pseudo Functions to Control Numerical Dispersion," *Soc. Pet. Eng. J.* (Aug. 1975) 269.
8. Camy, J.P. and Emanuel, A.S.: "Effect of Grid Size in the Compositional Simulation of CO_2 Injection," paper SPE 6894 presented at the 52nd Annual Technical Conference and Exhibition, Denver, Oct. 9–12, 1977.

9. Breit, V. and Grane, D.: "Scaling of Flow Parameters for Miscible Gas Flood Simulation Studies," paper SPE/DOE 9804 presented at the 2nd Joint Symposium EOR, Tulsa, April 5–8, 1981.
10. Coats, K.H., Dempsey, J.R., and Henderson, J.H.: "The Use of Vertical Equilibrium in Two-Dimensional Simulation of Three-Dimensional Reservoir Performance," *Soc. Pet. Eng. J.* (March 1971) 63.
11. Hearn, C.L.: "Simulation of Stratified Waterflooding by Pseudo Relative Permeability Curves," *J. Pet. Tech.* (July 1971) 805.
12. Thomas, G.W. and Thurnau, D.H.: "An Extension of Pseudo Function Concepts," forthcoming.
13. van Poollen, H.K., Breitenbach, E.A., and Thurnau, D.H.: "Treatment of Individual Wells and Grids in Reservoir Modeling," *Soc. Pet. Eng. J.* (Dec. 1968) 341.
14. Jacks, H.H., Smith, O.J., and Mattax, C.C.: "The Modeling of a Three-Dimensional Reservoir with a Two-Dimensional Reservoir Simulator—The Use of Dynamic Pseudo Functions," *Soc. Pet. Eng. J.* (June 1973) 175.
15. Stright, D.S., Jr.: "Grand Forks—Modeling a Three-Dimensional Reservoir with Two-Dimensional Reservoir Simulators," *J. Can Pet. Tech.* (Oct.–Dec. 1973) 1248.
16. Emanuel, A.S. and Cook, G.W.: "Pseudo-Relative Permeability for Well Modeling," *Soc. Pet. Eng. J.* (Feb. 1974) 7.
17. Chappelear, J.E. and Hirasaki, G.J.: "A Model of Oil-Water Coning for Two-Dimensional, Areal Reservoir Simulation," *Soc. Pet. Eng. J.* (April 1976) 65.
18. Woods, E.G. and Khurana, A.K.: "Pseudo Functions for Water Coning in a Three-Dimensional Reservoir Simulator," *Soc. Pet. Eng. J.* (Aug. 1977) 251.
19. Fagin, R.G., Irby, T.L., and McCord, D.L.: "Application of Two-Dimensional Network Models to Field Problems," paper SPE 725 presented at the 38th Annual Meeting, New Orleans, Oct. 6–9, 1963.
20. Coats, K.H.: "A Treatment of the Gas Percolation Problem in Simulation of Three-Dimensional Three-Phase Flow in Reservoirs," *Trans.*, AIME (1968) **243**, 413.
21. McCreary, J.G.: "A Simple Method for Controlling Gas Percolation in Numerical Simulation of Solution Gas Drive Reservoirs," *Trans.*, AIME (1971) **251**, 85.
22. Steffensen, R.J. and Sheffield, M.: "Reservoir Simulation of a Collapsing Gas Saturation Requiring Areal Variation in Bubble-Point Pressure," paper SPE 4275 presented at the 3rd Symposium on Numerical Simulation of Reservoir Performance, Houston, 1973.
23. Kazemi, H.: "A Reservoir Simulator for Studying Productivity Variation and Transient Behavior of a Well in a Reservoir Undergoing Gas Evolution," *Trans.*, AIME (1975) **259**, 1401.
24. Thomas, L.K., Lumpkin, W.B., and Reheis, G.M.: "Reservoir Simulation of Variable Bubble-Point Problems," *Trans.*, AIME (1976) **261**, 10.
25. Stright, D.H., Jr., Aziz, K., Settari, A., and Starratt, F.: "Carbon Dioxide Injection into Bottom—Water Undersaturated Viscous Oil Reservoirs," *J. Pet. Tech.* (Oct. 1977) 1248.
26. Thurnau, D.H. and Thomas, G.W.: "Special Concepts in Black Oil Reservoir Simulators," forthcoming.

Nomenclature

English

A, B, C, etc.	Denote arbitrary matrices.
A	Cross-sectional area.
a, b, c, etc.	Denote arbitrary vectors.
a_i, b_i, c_i, etc.	Arbitrary coefficients or components of a vector.
a_{ij}, b_{ij}, c_{ij}, etc.	Matrix elements.
B_l	Formation volume factor of phase l.
b_l	Shrinkage factor of phase $l = 1/B_l$.
C_{ij}	Cofactor of a matrix.
C_{il}	Mass fraction of component i in phase l.
c	Arbitrary constant, chapter 2. Compressibility of a volume occupied by a single-phase fluid, psi^{-1}, chapter 6.
c_b	Bulk rock volume compressibility, psi^{-1}.
c_l	Compressibility of fluid phase l, psi^{-1}.
c_p	Rock pore volume compressibility, psi^{-1}.
c_r	Rock compressibility in terms of porosity, psi^{-1}.
c_s	Solid compressibility of rock, psi^{-1}.
d	Arbitrary distance below or above a datum, ft.
d_k	Depth to an arbitrary layer, k, ft.
d_{kU}	Depth to the uppermost layer open to flow into a well, ft.
d_φ	Average pore diameter of a reference volume.
E	Set of nodes from the graph of a matrix.
E_n	Euclidean space, n-dimensional.
$\mathbf{e}^n$	Error vector at time level n.
f_{lk}	Fractional flow of phase l into a well from layer k.
$G(\mathbf{A})$	Directed graph of matrix $\mathbf{A}$.
$G(X, E), G$	Undirected graph of a matrix.
$G'(X', E'), G'$	Undirected subgraph of a matrix.

g/g_c	Ratio of the gravitational constant to that at sealevel.
$\tilde{g}$	Mass rate (source/sink) term for single-phase flow, units arbitrary.
$\tilde{g}_l$	Mass rate (source/sink) term for flow of phase l.
H_k	Normalized iteration parameter.
h	Reservoir thickness, ft.
$\mathbf{I}$	Identity matrix.
$\mathbf{i}$	Unit x-direction cartesian vector.
$\mathbf{j}$	Unit y-direction cartesian vector.
K	Total number of ADI iteration parameters.
K_i	Vapor-liquid equilibrium K-value for component i.
K_{igo}	Equilibrium K-value for component i partitioning between gas and oil.
K_{igw}	Equilibrium K-value for component i partitioning between gas and water.
K_v	Ratio of the volume of gas flow from a grid block in a time step to the gas in place within the block at the beginning of the time step.
k_L	Lowermost layer index open to flow into a well, ft.
k_l	Effective permeability of phase l, md.
k_{rl}	Relative permeability to phase l.
k_{row}	Relative permeability to oil, water-oil system.
k_{rog}	Relative permeability to oil, gas-oil system.
k^*_{row}, k^*_{ro}	End-point oil relative permeability, $k_{row}(S_{wr})$, chapters 3 and 8.
k^*_{rw}	End-point water relative permeability, $k_{rw}(1 - S_{or})$, chapter 8.
k'_{rg}	Derivative of gas relative permeability with respect to gas saturation.
k'_{row}	Derivative of oil relative permeability with respect to water.
k'_{rog}	Derivative of oil relative permeability with respect to gas.
$\bar{k}_{rl}$	Pseudo relative permeability to phase l.
k_U	Uppermost layer index open to flow into a well.
k_x	Permeability in the cartesian x-direction, md.
k_y	Permeability in the cartesian y-direction, md.
k_z	Permeability in the cartesian z-direction, md.
k_{xy}	Premeability in the cartesian xy-plane, md.
$[k]$	Permeability tensor, md.
k_∞	Constant in Klinkenburg permeability, md.
$\mathbf{k}$	Unit z-direction cartesian vector.
$\mathbf{L}$	Lower triangular matrix, chapter 2. Characteristic length in a porous medium wherein porosity varies significantly, ft. and moles of hydrocarbon liquid, chapter 3. Arbitrary distance, ft., chapters 4 & 5. Differential operator, chapter 5.
$\tilde{\mathbf{L}}$	Lower triangular matrix in the strongly implicit procedure.
L_Δ	Finite difference operator.
l	Characteristic distance across a reference volume, ft.
l_{ij}	Element of a lower triangular matrix.
$\mathbf{M}$	Incidence matrix, chapter 6. A variable in the leap-frog method, chapter 7.
$\mathbf{M}_{GS}$	Gauss-Seidel iteration matrix.
M_i	Molecular weight of component i.
M_{ij}	Minor of a matrix.
$\mathbf{M}_J$	Point Jacobi iteration matrix.
M_l	Finite difference operator for phase l, chapter 7.

Nomenclature

$\mathbf{M}_\omega$	Point SOR iteration matrix.
$m(p)$	Real gas pseudo pressure function.
N	A variable in the leap frog method, chapter 7.
N_l	Finite difference operator for phase l, chapter 7.
N_x	Number of grid blocks in the cartesian x-direction.
N_y	Number of grid blocks in the cartesian y-direction.
N_z	Number of grid blocks in the cartesian z-direction.
n_t	Total number of moles.
$\mathbf{n}$	Unit vector at a point P in the direction of the outward drawn normal to a surface.
$\mathbf{0}$	Null or zero vector.
P	A point in some region, chapter 2. A variable in the leap frog method, chapter 7. A permutation matrix, Appendix.
P_{cwo}	Capillary pressure, water-oil system, psi.
P_{cgo}	Capillary pressure, gas-oil system, psi.
P'_{cwo}	Derivative of water-oil capillary pressure with respect to water saturation, psi.
P'_{cgo}	Derivative of gas-oil capillary pressure with respect to gas saturation, psi.
$\tilde{P}_c(x,y)$	Pseudo capillary pressure in the cartesian x-y plane, psi.
PI	Productivity index. STB/day-psi.
p	An arbitrary pressure, psi.
p_l	Pressure in phase l, psi.
p_{nw}	Nonwetting phase pressure, psi.
p_o	Pressure at a datum, psi.
p_s	Saturation or bubble-point pressure, psi.
p_w	Wetting phase or water phase pressure, psi.
p_{wf}	Actual flowing bottomhole pressure, psi.
$\bar{p}_{wf}$	Flowing bottomhole pressure constraint, psi.
Q_g	Gas flow from a grid block over a time step, Mcf, chapter 8.
Q_l	Volumetric source/sink term for phase l, Vol./Vol.-unit time.
q_l	Volumetric flow rate for phase l, STB/day, or Mcf/day.
q_{lk}	Volumetric flow rate of phase l from layer k, STB/day or Mcf/day.
$\bar{q}_l$	A desired or specified well rate for phase l, STB/day or Mcf/day.
$\bar{q}_t$	A specified total fluid rate on a well, STB/day or Mcf/day.
$\mathbf{q}$	Mass flux, mass/unit area–unit time.
$\tilde{q}$	$-(B_w q_w + B_o q_o)$, chapter 7.
R	A region or reservoir, chapters 2 & 6. A variable in the leap-frog method, chapter 7.
R_{ij}	Residual at a point for a 2-D areal, single-phase flow problem, chapter 6, STB.
R_l^k	Residual of phase l at iteration level k, STB or Mcf.
R_s	Solution gas-oil ratio, Mcf/STB.
r_e	Radius of drainage of a well, ft.
r_w	Wellbore radius, ft.
S	Source/sink term $\equiv \pm\tilde{g}$.
S^*	Source/sink term, $= S\mu/\sigma k$, chapter 4.
S_c	A surface in cartesian three-space.
S_l	Saturation of phase l.
S_{wr}	Irreducible water saturation.
S_{or}	Irreducible oil saturation.

S_1	Storage for normal grid ordering.
S_4	Storage for alternate diagonal ordering.
S'	$\left(\dfrac{dP_{cwo}}{dS_w}\right)^{-1}$, psi^{-1}.
$\tilde{S}_l$	Pseudo saturation of phase l.
s	Integration variable over a surface, where ds is an element of surface, chapter 2. Skin factor, chapter 8.
T	Temperature, °R.
T_l	Transmissibility of phase l.
T_s	Temperature at standard conditions, °R.
$T(X, E)$	A tree, Appendix.
$\tilde{T}$	$T_o + T_w$, chapter 7.
t	Dimensionless time, chapter 5.
U	Upper triangular matrix.
$\tilde{U}$	Upper triangular matrix in the strongly implicit procedure.
u	Normalized pressure.
u_1, u_2	Arbitrary functions, Appendix.
u	Exact solution to a difference equation.
$u*$	Approximate solution to a difference equation.
u_{ij}	Elements of an upper triangular matrix.
V	Moles of vapour.
V_b	Bulk volume, ft^3.
V_p	Pore volume at a datum, ft^3.
V_s	Grain volume, ft^3.
v_o	Reference volume, ft^3.
v	An arbitrary vector, chapter 2.
v_i	Superficial velocity, chapter 4. Eigenvector.
v_1, v_2, v_3, etc.	Scalar components of v.
WI	Wellbore index, STB-cp/day-psi.
w	Relative permeability weighting factor.
X	Distance, ft, chapter 5. Set of edges, Appendix.
x	Cartesian coordinate, chapter 2. Dimensionless distance, chapter 5.
x_i	Mole fraction of component i in the liquid phase.
$x_1, x_2, x_3 \ldots$	Arbitrary unknowns.
y	Cartesian coordinate.
y_i	Mole fraction of component i in the vapor phase.
y	A vector in LU decomposition.
z	Cartesian coordinate, chapter 2. Gas deviation factor, chapter 3.
z_c	Oil-water contact at some time t, ft.
z_{ci}	Original oil-water contact, ft.
z_i	Overall mole fraction of component i.

Greek

α	Scalar quantity, chapter 2. Hydraulic diffusivity, $k/\varphi\mu c$, chapters 4 and 6. A real or complex number, chapter 5. Pore volume-compressibility product, $c\Delta x_i \Delta y_j \Delta z_k \varphi$, chapter 6. Unit conversion factor, chapter 7.
β	Scalar quantity, chapter 2. A factor, $\varphi\mu c\Delta x_i \Delta y_j/k\Delta t$, chapter 6.
γ_l	Specific density of phase l, psi/ft, chapter 4. A factor, $\Delta t/\Delta x^2$, chapter 5.

Nomenclature

Δ	Finite difference operator or incremental change in a variable, e.g., $\Delta x = x_1 - x_2$.
$\Delta_t, \bar{\delta}$	Change over a time step, e.g., $\Delta_t X = \bar{\delta} X = X^{n+1} - X^n$.
δ	Incremental change in a variable or change over an iteration, e.g., $\delta \gamma = \gamma_w - \gamma_o$ or $\delta x = x^{k+1} - x^k$.
$\epsilon, \epsilon_1, \epsilon_2$	Closure tolerances.
ϵ_i	Component of total error for a 1-D problem.
ϵ_{ij}	Error component for a 2-D problem.
ϵ_o	Closure tolerance for the oil equation.
ϵ_w	Closure tolerance for the water equation.
ζ	A factor in von Neumann stability analysis, $\zeta = \exp(\alpha \Delta t)$, for α a real or complex number.
θ	Dip angle, degrees.
θ_c	Contact angle, degrees.
λ, λ_i	Eigenvalue of a matrix.
λ_l	Phase mobility of phase l; subsurface conditions, k_{rl}/μ_l; surface conditions, $k_{rl} b_l / \mu_l$.
μ	Viscosity of a single-phase fluid, cp.
μ_l	Viscosity of phase l, cp.
ν	Integration variable over a volume where dv is an element of volume.
ξ	Proportionality parameter.
π	Transcendental number, 3.1415 . . .
ρ	Density of a single-phase fluid, lbs/ft³.
ρ_l	Density of phase l, lbs/ft³.
ρ_{ls}	Density of phase l at standard conditions, lbs/ft³.
$\rho(A)$	Spectral radius of matrix A.
$\rho(M_J)$	Spectral radius of the point Jacobi iteration matrix.
ΣT	Sums of the transmissibilities over the faces of a grid block.
σ	Spectral radius of the point Jacobi iteration matrix, $\sigma = \rho(M_J)$.
σ_k	ADI iteration parameter.
σ_1, σ_K	Minimum and maximum ADI iteration parameters, respectively.
$\sigma_{os}, \sigma_{ws}, \sigma_{ow}$	Interfacial energies for oil-solid, water-solid and oil-water, respectively.
τ	Real time, hours or days.
Υ	Truncation error.
Φ	A scalar function in cartesian coordinates.
Φ_h	Hubbert potential, psi.
Φ_l	Fluid potential of phase l, psi.
φ, ϕ	Porosity.
ϕ_o	Porosity at a datum.
ψ	Constant or time-invariant part of transmissibility, $RB\text{-}cp/\text{day-psi}$.
ω	Iteration parameter for SIP or SOR, chapter 6. A number, 0 or 1, chapter 7.
ω_o	Optimum SOR iteration parameter.

Mathematical Symbols

∇	Del or nabla operator.
Σ	Summation.
$\int_a^b$	The integral taken between the values of a and b of the variable.

$>$	Greater than.
$\gg$	Much greater than.
$<$	Less than.
$\geq$	Greater than or equal to.
$\leq$	Less than or equal to.
$\neq$	Does not equal.
$\approx$	Is approximately equal to.
$\equiv$	Is identical to or defined by.
$\therefore$	Therefore.
$\exists$	There exists.
$\ni$	Such that.
$\forall$	For every.
ε, ϵ	Is an element of or belongs to.
$\Rightarrow$	Implies.
$!$	Factorial, e.g., $5! = 1 \cdot 2 \cdot 3 \cdot 4 \cdot 5$.
$[a, b]$	The closed interval, $a \leq x \leq b$.
$\{a_i\}_i^n$	The set of elements $a_1, a_2, \ldots, a_n$.
$\|z\|$	Absolute value of z.
$\|\mathbf{v}\|$	The norm or magnitude of the vector $\mathbf{v}$.
$\|\mathbf{A}\|$	The determinant of matrix $\mathbf{A}$.
$\ldots$	And so on.
∂R	Boundary of a region R.
$\dfrac{d}{ds}$	Ordinary derivative operator with respect to s.
$\dfrac{\partial}{\partial s}$	Partial derivative operator with respect to s.

Abbreviations

ADI	Alternating Direction Implicit Procedure.
AIM	Adaptive Implicit Method.
cp	Centipoise.
DKR	Dupont, Kendall, Rachford.
D4	Alternate diagonal ordering.
ft	Feet.
ft³	Cubic feet.
GIP	Gas in place.
IADI	Iterative alternating direction implicit procedure.
IMPES	Implicit pressure, explicit saturation.
LSOR	Line successive overrelaxation.
lbs	Pounds.
MBE	Material balance equation.
Mcf	Thousand cubic feet.
md	Millidarcies.
PI	Productivity index.
RB	Reservoir barrels.
SIP	Strongly implicit procedure.
SOR	Successive overrelaxation.
STB	Stock tank barrels.
VE	Vertical equilibrium.
psi	Pounds per square foot.
1-D	One-dimensional in space.
2-D	Two-dimensional in space.
3-D	Three-dimensional in space.
2LSOR	Two line successive overrelaxation.

Appendix

Logic, like whiskey, loses its beneficial effect when taken in too large quantities.

Lord Dunsany

We collect here some mathematical concepts that will be useful in our applications to reservoir simulation. In particular a number of theorems are given, some with proofs, and some without. The reader is urged to go through the proofs in detail to better his understanding. When a proof is not given, it is because it is either lengthy and complex, or requires a background that is beyond the scope of our purposes.

A.1 Big "0" Notation

Let f and g be real valued functions defined on a set S of real numbers and assume g is nonnegative. We write

$$f(x) = 0(g(x)), x \in S$$

if $\exists$ a positive number $M \ni |f(x)| \leq M\, g(x)\; \forall\; x \in S$. The actual value of M is not important. We need only know that some positive constant, M, exists satisfying the condition.

Example: Let $g(x) = 1;\; f(x) = \sin x$.

Then $\sin x = 0\,(1)$ because

$|\sin x| \leq 1$ where here $M = 1$.

A.2 Functions of Class C^n

A function f, defined on some domain, D, is said to be of class C^n, if and only if, it and its partial derivatives up to and including the nth derivative exist and are continuous in D.

Example 1: Let $D \equiv (0, \pi)$, $f(x) = \sin x$. Then $f(x) \, \varepsilon \, C^\infty$ since f and all derivatives of f exist and are continuous in D.

Example 2: Let $D \equiv (-\infty, \infty)$ $f(x) = \begin{cases} \sin(1/x), & x \neq 0 \\ 0, & x = 0. \end{cases}$

Then $f(x) \, \varepsilon \, C^0$ since $f'(x) = -(1/x^2) \cos(1/x)$ but $f'(0)$ does not exist.

A.3 First Mean Value Theorem for Integrals

Let $f(x)$ be bounded and continuous $\forall \, x \, \varepsilon [a, b]$. This $\Rightarrow \exists$ numbers N and $M \ni N \leq |f(x)| \leq M \, \forall \, x \, \varepsilon [a, b]$ and $f(x)$ is continuous in $[N, M]$. Then $\exists$ a number $\zeta \, \varepsilon (a, b) \ni \int_a^b f(x) \, dx = (b-a) f(\zeta)$ where $\zeta = a + (b-a)\theta$, $0 < \theta < 1$. (See Whittaker & Watson[1] for proof.)

A.4 Leibniz' Rule for Integrals

Let $f(x,y)$ be continuous having a continuous partial derivative $f_x(x,y)$ in some interval $D \equiv [a(x), b(x)]$. If $g(x) = \int_{a(x)}^{b(x)} f(x,y) dy$, then

$$g'(x) = f[x, b(x)] b'(x) - f[x, a(x)] a'(x) + \int_{a(x)}^{b(x)} f_x(x,y) dy. \tag{A.1}$$

Proof:

$$g'(x) = \lim_{\Delta x \to 0} \frac{1}{\Delta x} \left\{ \int_{a(x+\Delta x)}^{b(x+\Delta x)} f(x+\Delta x, y) dy - \int_{a(x)}^{b(x)} f(x,y) dy \right\}$$

$$= \lim_{\Delta x \to 0} \frac{1}{\Delta x} \left\{ \left[\int_{a(x+\Delta x)}^{b(x+\Delta x)} f(x+\Delta x, y) dy - \int_{a(x+\Delta x)}^{b(x)} f(x+\Delta x, y) dy \right] \right.$$

$$+ \left[\int_{a(x+\Delta x)}^{b(x)} f(x+\Delta x,y)dy - \int_{a(x)}^{b(x)} f(x+\Delta x,y)dy\right]$$

$$+ \left[\int_{a(x)}^{b(x)} f(x+\Delta x,y)dy - \int_{a(x)}^{b(x)} f(x,y)dy\right]\bigg\}$$

$$= \lim_{\Delta x \to 0} \frac{1}{\Delta x} \bigg\{ \left[\int_{b(x)}^{b(x+\Delta x)} f(x+\Delta x,y)dy - \int_{a(x)}^{a(x+\Delta x)} f(x+\Delta x,y)dy\right.$$

$$+ \int_{a(x)}^{b(x)} [f(x+\Delta x,y) - f(x,y)]dy\bigg\}$$

$$= \lim_{\Delta x \to 0} \bigg\{ \frac{b(x+\Delta x) - b(x)}{\Delta x} f(x,\zeta_1) - \frac{a(x+\Delta x) - a(x)}{\Delta x} f(x,\zeta_2)$$

$$+ \int_{a(x)}^{b(x)} \frac{f(x+\Delta x,y) - f(x,y)}{\Delta x} dy\bigg\}$$

from the first mean value theorem where $\zeta_1 = b(x) + \theta_1[b(x+\Delta x) - b(x)]$ and $\zeta_2 = a(x) + \theta_2[a(x+\Delta x) - a(x)]$. In the limit we obtain Eq. A.1.

A.5 Classification of Partial Differential Equations

Most equations of mathematical physics are of the form

$$a(x,y)u_{xx} + 2b(x,y)u_{xy} + c(x,y)u_{yy} = G(u, u_x, u_y, x, y). \quad (A.2)$$

We define a differential operator L as

$$L = a(x,y)\frac{\partial^2}{\partial x^2} + 2b(x,y)\frac{\partial^2}{\partial x \partial y} + c(x,y)\frac{\partial^2}{\partial y^2} - G$$

such that Eq. A.2 becomes

$$L\{u\} = 0.$$

A.5.1 Linearity

An operator is said to be linear, if given two functions u_1 and u_2 that are sufficiently differentiable, then

$$L\{\alpha_1 u_1 + \alpha_2 u_2\} = \alpha_1 L\{u_1\} + \alpha_2 L\{u_2\}$$

where α_1 and α_2 are constants otherwise L is nonlinear. If L is linear, then it can be shown by an extension of the above definition that

$$L\left\{\sum_{i=1}^{n} \alpha_i u_i\right\} = \sum_{i=1}^{n} \alpha_i L\{u_i\}$$

where the sets

$$\{\alpha_i\}_{i=1}^{n} \text{ and } \{u_i\}_{i=1}^{n}$$

are scalars and functions respectively.

Let $g(x,y)$ and $h(x,y)$ be two functions belonging to C^2 on D so that the first and second derivatives exist. We define the Jacobian to be the determinant:

$$J = \begin{vmatrix} g_x & g_y \\ h_x & h_y \end{vmatrix}$$
$$= g_x h_y - h_x g_y.$$

If $J \neq 0$ for the two functions $g(x,y)$ and $h(x,y)$ then inverse functions exist; i.e., we can find x and y in terms of g and h.

$$x = x(g,h)$$
$$y = y(g,h)$$

If so, we make a change of variable in Eq. A.2 to give

$$a u_{xx} + 2b u_{xy} + c u_{yy} = \alpha u_{gg} + 2\beta u_{gh} + \gamma u_{hh} + \text{lower order terms} \quad (A.3)$$

where

$$\alpha \equiv a g_x^2 + 2b g_x g_y + c g_y^2$$
$$\beta \equiv a g_x h_x + b(g_x h_y + g_y h_x) + c g_y h_y$$
$$\gamma \equiv a h_x^2 + 2b h_x h_y + c h_y^2$$

Thus, the change of variable in Eq. A.2 produces Eq. A.3 with the relationship in the coefficients

$$\beta^2 - \alpha\gamma = (b^2 - ac)(g_x h_y - g_y h_x)^2. \quad (A.4)$$

(Proof of this is left as an exercise for the reader.) Note the sign of $(b^2 - ac)$ remains invariant under the variable transformation. Thus, $(b^2 - ac)$ is a fundamental property of the partial differential equation called the *discriminant*. Partial differential equations are classified as elliptic, hyperbolic or parabolic based on the following criteria:

$$b^2 - ac < 0, \; L\{u\} \text{ is elliptic}$$
$$b^2 - ac > 0, \; L\{u\} \text{ is hyperbolic}$$
$$b^2 - ac = 0, \; L\{u\} \text{ is parabolic}$$

A.5.2 Canonical Forms

When an equation is not readily expressed in the form of Eq. A.2 or Eq. A.3, we nevertheless can bring it into a special normal or *canonical form* by an appropriate change in the independent variable. This results in

or

$$\left. \begin{array}{l} \dfrac{\partial^2 u}{\partial g \partial h} = F(u, g, h, u_g, u_h) \\[6pt] \dfrac{\partial^2 u}{\partial g^2} - \dfrac{\partial^2 u}{\partial h^2} = H(u, g, h, u_g, u_h) \end{array} \right\} \text{ for hyperbolic eqs.,}$$

$$\dfrac{\partial^2 u}{\partial g^2} + \dfrac{\partial^2 u}{\partial h^2} = F(u, g, h, u_g, u_h) \quad \text{for elliptic eqs.,}$$

and

$$\dfrac{\partial^2 u}{\partial g^2} = F(u, g, h, u_g, u_h) \quad \text{for parabolic eqs.}$$

Linear equations of second order involving more than two independent variables can also be put in canonical form. However, in some cases, it is not possible to find a transformation that yields a canonical form in the entire neighbourhood of the point at which the equations are defined, but rather, only for the point itself. If, by a suitable transformation, an equation can be reduced at this point to the form

$$\sum_{k=1}^{n} \frac{\partial^2 u}{\partial x_k^2} + \ldots = 0, \tag{A.5a}$$

$$\text{or} \quad \sum_{k=1}^{n-1} \frac{\partial^2 u}{\partial x_k^2} - \frac{\partial^2 u}{\partial x_n^2} + \ldots = 0, \tag{A.5b}$$

$$\text{or} \quad \sum_{k=1}^{n} \frac{\partial^2 u}{\partial x_k^2} + a_1 \frac{\partial u}{\partial x_1} + \ldots = 0, \tag{A.5c}$$

where n is the number of independent variables, then the equation is elliptic, hyperbolic, or parabolic, respectively. In Eqs. A.5a–A.5c we show only the terms involving second order derivatives except for Eq. A.5c where a_1 must be nonzero.

A.5.3 Solution of Partial Differential Equations

The solution of a partial differential equation is dependent upon the nature of the equation itself (or classification) and some additional data. The additional data must be compatible with the nature of the equation. When it is, we say the problem is *well posed* (in the sense of Hadamard).

For elliptic equations prescribed on a region R having a boundary ∂R, a well-posed problem requires the following:

(1) ∂R must be closed and finite.
(2) Values of the dependent variable or its gradient or both must be specified at all points on ∂R.

For hyperbolic and parabolic equations on some region R there is always a "time-like" derivative involved with respect to some independent variable, t say, where $0 < t < \infty$. A well-posed problem for these types of equations requires the following:

(1) R must be open, i.e., ∂R is not a closed boundary.
(2) The auxiliary data consists of:
   (a) Boundary conditions on the spatial part of the problem.
   (b) Initial conditions on the part involving the time-like derivative.

A.6 Matrix Methods

Here we collect a number of results from matrix theory that are frequently important in determining whether or not a given numerical procedure is convergent and stable. We also develop the procedure for LU-decomposition which is important in solving some matrix problems.

A.6.1 Rank of a Matrix

A *submatrix* of matrix **A** is defined as either **A** itself or any array remaining after certain rows or columns (lines) are deleted from **A**. The square submatrices of **A** are particularly useful. The determinant of order r of **A** is the determinant of an $r \times r$ submatrix of **A**. A matrix is said to be of rank r denoted by $r(\mathbf{A})$ if and only if it has at least one nonsingular $r \times r$ submatrix, but no nonsingular submatrix of order $>r$. A matrix has a rank of zero if and only if it is the null matrix, i.e., all its elements are zero.

Appendix

The rank of **A** is also the maximum number of linearly independent row (or column) vectors in **A**.

A.6.2 Linear Independence of Eigenvectors

Theorem: If the eigenvalues of **A** are distinct, i.e., if each has multiplicity of unity, then their corresponding eigenvectors are linearly independent.

Proof: We consider the case of three distinct eigenvalues, λ_1, λ_2 and λ_3 and leave the more general case as an exercise. Let the associated eigenvectors be x_1, x_2, x_3 of matrix **A**. Then

$$\mathbf{A}x_1 = \lambda_1 x_1$$
$$\mathbf{A}x_2 = \lambda_2 x_2$$
$$\mathbf{A}x_3 = \lambda_3 x_3$$

where $\lambda_1 \neq \lambda_2 \neq \lambda_3 \neq \lambda_1$. To show that $\{x_i\}_{i=1}^{3}$ are linearly independent we need only show that

$$\sum_{i=1}^{3} c_i x_i = 0 \Rightarrow c_i = 0, \ i = 1, 2, 3. \tag{A.6}$$

Premultiply Eq. A.6 by **A**,

$$\sum_{i=1}^{3} c_i \mathbf{A} x_i = 0 \Rightarrow \sum_{i=1}^{3} c_i \lambda_i x_i = 0. \tag{A.7}$$

Repeating this again we have

$$\sum_{i=1}^{3} c_i \lambda_i^2 x_i = 0. \tag{A.8}$$

Eqs. A.6–A.8 can be put in the matrix form

$$\begin{bmatrix} 1 & 1 & 1 \\ \lambda_1 & \lambda_2 & \lambda_3 \\ \lambda_1^2 & \lambda_2^2 & \lambda_3^2 \end{bmatrix} \begin{bmatrix} c_1 x_1 \\ c_2 x_2 \\ c_3 x_3 \end{bmatrix} = \begin{bmatrix} 0 \\ 0 \\ 0 \end{bmatrix}. \tag{A.9}$$

Define $\mathbf{B} \equiv \begin{bmatrix} 1 & 1 & 1 \\ \lambda_1 & \lambda_2 & \lambda_3 \\ \lambda_1^2 & \lambda_2^2 & \lambda_3^2 \end{bmatrix}$ (a Vandermonde matrix) where $|\mathbf{B}| = (\lambda_1 - \lambda_2)(\lambda_3 - \lambda_2)(\lambda_3 - \lambda_1)$. Since $\lambda_i \neq \lambda_j$ $i \neq j$, $i,j = 1,2,3$ then $|\mathbf{B}| \neq 0$, $\therefore \mathbf{B}^{-1}$ exists and we write Eq. A.9 as

$$\begin{bmatrix} c_1 x_1 \\ c_2 x_2 \\ c_3 x_3 \end{bmatrix} = \mathbf{B}^{-1} \begin{bmatrix} 0 \\ 0 \\ 0 \end{bmatrix} = \begin{bmatrix} 0 \\ 0 \\ 0 \end{bmatrix}. \qquad (A.10)$$

Since x_1, x_2 and x_3 are eigenvectors, they are nonzero which $\Rightarrow c_1 = c_2 = c_3 = 0$.

Theorem: If λ is an eigenvalue of multiplicity k of an $n \times n$ matrix $\mathbf{A}$, then the number of linearly independent eigenvectors of $\mathbf{A}$ associated with λ is $\rho = n - r(\mathbf{A} - \lambda \mathbf{I})$. Furthermore, $1 \leq \rho \leq k$. Notice, we can always find at least n linearly independent eigenvectors for every $n \times n$ matrix.

A.6.3 Polynomials of a Matrix

We define $\mathbf{A}.\mathbf{A}.\mathbf{A} \ldots \ldots \mathbf{A}$ (n times) $= \mathbf{A}^n$. We also define a matrix polynomial $\mathbf{B}$ where

$$\mathbf{B} = \sum_{k=0}^{n} \alpha_k \mathbf{A}^k$$

$$= \alpha_0 \mathbf{A}^0 + \alpha_1 \mathbf{A}^1 + \alpha_2 \mathbf{A}^2 + \alpha_3 \mathbf{A}^3 + \ldots \alpha_n \mathbf{A}^n$$

and $\mathbf{A}^0 = \mathbf{I}$.

Theorem: If $\mathbf{A}$ is $n \times n$ then we can show that every function of $\mathbf{A}$ is related to a polynomial of $\mathbf{A}$, e.g.,

$$\cos \mathbf{A} = \sum_{k=0}^{n} \alpha_k \mathbf{A}^k$$
$$= f(\mathbf{A}).$$

Theorem: If λ is an eigenvalue of $\mathbf{A}$ then $f(\lambda)$ is an eigenvalue of $f(\mathbf{A})$.

A.6.4 Gerschgorin's Theorem[2]

Let $\mathbf{A}$ be $n \times n$ and let the set $\{S_i\}_{i=1}^{n}$ be the sums of the moduli of the elements along the rows or columns of $\mathbf{A}$, where

$$S_m = \max_i \{S_i\}.$$

Appendix

Then the spectral radius of **A** is less than or equal to S_m, i.e.,

$$\rho(\mathbf{A}) \leq S_m.$$

Proof: Let λ_i be an eigenvalue of **A** and let the associated eigenvector be **X**. Then we have by definition:

$$\mathbf{AX} = \lambda_i \mathbf{X}$$

or

$$a_{11}x_1 + a_{12}x_2 + \ldots + a_{1n}x_n = \lambda_i x_1$$
$$a_{21}x_2 + a_{22}x_2 + \ldots + a_{2n}x_n = \lambda_i x_2$$
$$\vdots \qquad \vdots \qquad \qquad \vdots \qquad \vdots$$
$$a_{k1}x_1 + a_{k2}x_2 + \ldots \quad a_{kn}x_n = \lambda_i x_k.$$

At least one of the x_i's is the largest. Suppose it is x_k. Then $a_{k1}\dfrac{x_1}{x_k} + a_{k2}\dfrac{x_2}{x_k} + ,\ldots, + a_{kn}\dfrac{x_n}{x_k} = \lambda_k.$ Recall the triangle inequality:

$$|c| \leq |a| + |b|.$$

Applying an extension of this we get

$$\left| a_{k1}\frac{x_1}{x_k} \right| + \left| a_{k2}\frac{x_2}{x_k} \right| + \ldots + \left| a_{kn}\frac{x_n}{x_k} \right| \geq |\lambda_k|$$

$$|a_{k1}|\left|\frac{x_1}{x_k}\right| + |a_{k2}|\left|\frac{x_2}{x_k}\right| + \ldots + |a_{kn}|\left|\frac{x_n}{x_n}\right| \geq |\lambda_k|.$$

Since x_k is the largest then

$$\left|\frac{x_i}{x_k}\right| < 1 \quad \text{for every } i \text{ except } i = k.$$

Thus,

$$|a_{k1}| + |a_{k2}| + \ldots + |a_{kn}| \geq |\lambda_k|$$

or

$$S_m \geq \rho(\mathbf{A}).$$

Since the eigenvalues of **A** are the same as $\mathbf{A}^T$, the theorem also follows for column summing.

A.6.5 Brauer's Theorem[3]

Let $\{S_i\}$ be a set of numbers generated by summing the moduli of the elements along the rows or columns, where the element a_{kk} is excluded. Then let

$$S_m = \max_i \{S_i\}, \; i \neq k.$$

Brauer's theorem states that all the eigenvalues will fall within or on the boundary of the disc defined by

$$|\lambda - a_{kk}| \leq S_m.$$

Proof: Let λ_i be eigenvalue of **A**, and **X** be the corresponding eigenvector. Thus,

$$\mathbf{A}\mathbf{X} = \lambda_i \mathbf{X}.$$

Then

$$\begin{aligned}
a_{11}x_1 + a_{12}x_2 + \ldots + a_{1n}x_n &= \lambda_i x_1 \\
a_{21}x_2 + a_{22}x_2 + \ldots + a_{2n}x_n &= \lambda_i x_2 \\
\vdots \qquad \vdots \qquad \qquad \vdots & \quad \vdots \\
a_{k1}x_1 + a_{k2}x_2 + \ldots + a_{kn}x_n &= \lambda_i x_k \\
\vdots \qquad \vdots \qquad \qquad \vdots & \quad \vdots \\
a_{n1}x_1 + a_{n2}x_2 + \ldots + a_{nn}x_n &= \lambda_i x_n.
\end{aligned}$$

Let $x_k = \max_i [x_i]$. Thus, in the kth row, we have

$$a_{k1}\frac{x_1}{x_k} + a_{k2}\frac{x_2}{x_k} + \ldots + a_{kn}\frac{x_n}{x_k} = \lambda_i - a_{kk}$$

$$\left|a_{k1}\frac{x_1}{x_k}\right| + \left|a_{k2}\frac{x_2}{x_k}\right| + \ldots + \left|a_{kn}\frac{x_n}{x_k}\right| = |\lambda_i - a_{kk}|$$

$$|a_{k1}|\left|\frac{x_1}{x_k}\right| + |a_{k2}|\left|\frac{x_2}{x_k}\right| + \ldots + |a_{kn}|\left|\frac{x_n}{x_k}\right| = |\lambda_i - a_{kk}|$$

by the extended triangle inequality. Since $\left|\frac{x_i}{x_k}\right| < 1$, then setting $\frac{x_i}{x_k} = 1$, does not change the inequality. Therefore,

$$|a_{k1}| + |a_{k2}| + \cdots + |a_{kn}| \geq |\lambda_i - a_{kk}|$$

or

$$S_m \geq |\lambda_i - a_{kk}|.$$

A.6.6 Permutation Matrices

A permutation matrix is an $n \times n$ matrix with a single element equal to 1 and all others equal to zero in each row and each column; e.g.,

let $\mathbf{A} = \begin{bmatrix} a_{11} & a_{12} \\ a_{21} & a_{22} \end{bmatrix}$ and $\mathbf{P} = \begin{bmatrix} 0 & 1 \\ 1 & 0 \end{bmatrix}$

then, $\mathbf{PA} = \begin{bmatrix} 0 & 1 \\ 1 & 0 \end{bmatrix} \begin{bmatrix} a_{11} & a_{12} \\ a_{21} & a_{22} \end{bmatrix}$

$$\begin{bmatrix} a_{21} & a_{22} \\ a_{11} & a_{12} \end{bmatrix};$$

i.e., the two rows are interchanged or the elements are permuted.

A.6.7 Reducible Matrices

Let $\mathbf{A}$ be an $n \times n$ complex matrix (i.e., one whose elements are complex numbers). We say $\mathbf{A}$ is reducible if there exists a permutation matrix $\mathbf{P}$, such that

$$\mathbf{PAP}^T = \begin{bmatrix} \mathbf{A}_{11} & \mathbf{A}_{12} \\ 0 & \mathbf{A}_{22} \end{bmatrix}$$

where

$\mathbf{A}_{11} \equiv r \times r$ submatrix $r < n$
$\mathbf{A}_{22} \equiv (n-r) \times (n-r)$ submatrix
$0 \equiv (n-r) \times r$ null submatrix
$\mathbf{A}_{12} \equiv r \times (n-r)$ submatrix

Suppose we want to solve $\tilde{\mathbf{A}}\mathbf{x} = \mathbf{b}$ where $\tilde{\mathbf{A}} = \mathbf{PAP}^T$. We partition $\mathbf{X}$ as

$$\mathbf{X} = \begin{bmatrix} \mathbf{X}_1 \\ \mathbf{X}_2 \end{bmatrix}$$

where

$$\mathbf{X}_1 = \begin{bmatrix} x_1 \\ x_2 \\ \cdot \\ \cdot \\ x_r \end{bmatrix} \quad \text{and} \quad \mathbf{X}_2 = \begin{bmatrix} x_{r+1} \\ x_{r+2} \\ \cdot \\ \cdot \\ x_n \end{bmatrix}.$$

Thus, $\mathbf{AX} = \mathbf{b}$ can be written as

$$\begin{bmatrix} \mathbf{A}_{11} & \mathbf{A}_{12} \\ 0 & \mathbf{A}_{22} \end{bmatrix} \begin{bmatrix} \mathbf{X}_1 \\ \mathbf{X}_2 \end{bmatrix} = \begin{bmatrix} \mathbf{b}_1 \\ \mathbf{b}_2 \end{bmatrix}$$

or

$$\mathbf{A}_{11} \mathbf{X}_1 + \mathbf{A}_{12} \mathbf{X}_2 = \mathbf{b}_1$$
$$\mathbf{A}_{22} \mathbf{X}_2 = \mathbf{b}_2.$$

This reduces the original problem to two smaller matrix problems which are easier to handle. Moreover, computer storage is conserved since the last expression is used to first solve for $\mathbf{X}_2$ followed by a solution for $\mathbf{X}_1$.

A.6.8 Graph of a Matrix

Let $\mathbf{A}$ be an $n \times n$ complex matrix. We associate with $\mathbf{A}$ n-nodes, P_i, $i = 1, 2, \ldots, n$. For every nonzero value of a_{ij} in $\mathbf{A}$, we connect a directed path from P_i to node P_j, denoted by $P_i P_j$. The collection of all such directed paths is the directed graph of $\mathbf{A}$, denoted by $G(\mathbf{A})$. If for every ordered pair of nodes, there exists at least one directed path, then we say $G(\mathbf{A})$ is connected. For example,

if
$$\mathbf{A} = \begin{bmatrix} 1 & 1 \\ 0 & 1 \end{bmatrix}.$$

$\mathbf{A}$ here, is not connected, since there is no path from 2 to 1.

If
$$\mathbf{A} = \begin{bmatrix} 1 & 1 \\ 2 & 3 \end{bmatrix}$$

then $\mathbf{A}$ is connected. An irreducible matrix is one whose graph is connected. If it is not, then it is reducible.

Frequently, we are not concerned with directed paths in the graph of a matrix. If we do not designate a direction for the paths in a graph, we say it is *undirected*. More precisely, we define an undirected graph as a collection of a finite set of nodes, X say, together with a set of edges, E

Appendix

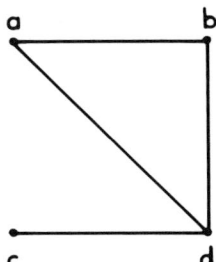

Fig. A.1 An Undirected Graph.

say, which are unordered pairs of distinct nodes. Instead of $G(A)$ we use the notation $G(X, E)$ to denote an undirected graph. For example, consider Fig. A.1. $X = \{a, b, c, d\}$ and $E = \{(a,b), (b,d), (a,d), (c,d)\}$. A subgraph $G'(X', E')$ is one where $X' \subseteq X$ and $E' \subseteq E$. For example, $X' = \{a, b, d\}$ and $E' = \{(a,b), (b,d), (c,d)\}$ defines a subgraph of the graph in Fig. A.1.

Observe that every connected directed graph (also called a connected *digraph*) has an associated undirected graph that is connected. A *cycle* or *loop* is a path that begins and ends at the same node. Another definition of a loop is:

(1) G' is connected.
(2) For every node in G', there are exactly two edges incident at it.

Here G' is any undirected subgraph of G. For example, consider Fig. A.2. The subgraph, G' having $X' = \{a, b, c, d\}$ and $E' = \{(a,b), (b,d), (c,d), (a,c)\}$ is a loop. However, the subgraph defined by $X' = \{a, b, e, f\}$ and $E' = \{(a,b), (a,e), (e,f)\}$ is not.

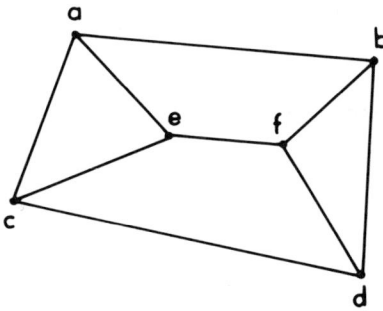

Fig. A.2 Example Undirected Graph.

A tree, $T(X,E)$ is a connected undirected graph with no loops. A subgraph G' of an undirected graph G is a tree if

(1) G' is connected.
(2) G' contains all nodes of G.
(3) G' has no loops.

Thus, the subgraph G' where $X' = \{a, b, c, d, e, f\}$ and $E' = \{(a,c), (a,b), (a,e), (e,f), (f,d)\}$ is a tree. If, on the other hand, for the same X' we have $E' = \{(a,c), (a,b), (a,e), (e,f), (f,b)\}$ then G' is not a tree.

Theorem: Matrices whose graphs are trees can always be ordered such that they produce no fill elements during Gaussian elimination.

A.6.9 Irreducible Diagonal Dominance

Let $\mathbf{A}$ be an $n \times n$ complex matrix, then $\mathbf{A}$ is diagonally dominant if

$$|a_{ii}| \geq \sum_{\substack{j=1 \\ i \neq j}}^{n} |a_{ij}|, 1 \leq i \leq n.$$

If strict inequality holds for every i, then $\mathbf{A}$ is *strictly diagonally dominant*. If $\mathbf{A}$ is irreducible and diagonally dominant with strict inequality for at least one value of i then we say $\mathbf{A}$ is *irreducibly diagonally dominant*.

Theorem: Let $\mathbf{A}$ be a $n \times n$ complex matrix, then $\mathbf{A}$ is non-singular if (a) $\mathbf{A}$ is strictly diagonally dominant or (b) $\mathbf{A}$ is irreducibly diagonally dominant.
Note: A matrix can be *irreducible* and *diagonally dominant* but still not be *irreducibly diagonally dominant* since for irreducible diagonal dominance we require strict inequality for at least one value of i.

A.6.10 LU Decomposition

This technique is valid for any square matrix, sparse or dense. Consider the matrix problem,

$$\mathbf{A}\mathbf{x} = \mathbf{b}. \qquad (A.11)$$

Let $\mathbf{A} = \mathbf{L}\mathbf{U}$ where $\mathbf{L} = [l_{ij}]$ and $l_{ij} = 0$ for $i < j$, $\mathbf{U} = [u_{ij}]$ and $u_{ij} = 0$ for $i > j$. $\mathbf{L}$ is called a lower *triangular matrix*. The elements on the main diagonal of **U** are all unity. **U** is called a *unit upper triangular matrix*. The structures of **L** and **U** are as follows:

$$L = \begin{bmatrix} l_{11} & & & \\ l_{21} & l_{22} & & \\ \cdot & & \cdot & \\ \cdot & & & \cdot \\ l_{n1} & \cdots & & l_{nn} \end{bmatrix}; U = \begin{bmatrix} 1 & u_{12} & u_{13} & \cdots & u_{1n} \\ & 1 & u_{23} & \cdots & u_{2n} \\ & & \cdot & & \cdot \\ & & & \cdot & \cdot \\ & & & & 1 \end{bmatrix}.$$

Substituting for A in Eq. A.11 we get $\mathbf{LUx} = \mathbf{b}$. Let $\mathbf{y} = \mathbf{Ux}$ thus, $\mathbf{Ly} = \mathbf{b}$. Assuming we know **L** we can find **y**. This is called the *forward solution*. We then use the definition of **y** to find **x**. This is a *backward solution*. Thus, we have the following three steps:

(1) Factorization
(2) Forward solution
(3) Backward solution

(1) Factorization:

$$\mathbf{A} = \mathbf{LU}$$

$$a_{ij} = \sum_{k=1}^{n} l_{ik} u_{kj}$$

$$= \sum_{k=1}^{j-1} l_{ik} u_{kj} + l_{ij} u_{jj} + \sum_{k=j+1}^{n} l_{ik} \overset{0}{u_{kj}}$$

since $u_{kj} = 0$ for $k > j$. Thus,

$$l_{ij} = a_{ij} - \sum_{k=1}^{j-1} l_{ik} u_{kj}. \tag{A.12}$$

Expanding column-wise we get

$$a_{ij} = \sum_{k=1}^{i-1} l_{ik} u_{kj} + l_{ii} u_{ij} + \sum_{k=i+1}^{n} l_{ik} \overset{0}{u_{ki}}.$$

Therefore, $$u_{ij} = \left(a_{ij} - \sum_{k=1}^{i-1} l_{ik} u_{kj} \right) \Big/ l_{ii}. \tag{A.13}$$

(2) Forward solution:

$$Ly = b$$

$$b_i = \sum_{k=1}^{n} l_{ik} y_k$$

$$= \sum_{k=1}^{i-1} l_{ik} y_k + l_{ii} y_i + \sum_{k=i+1}^{n} l_{ik} \overset{o}{y_k}$$

Consequently,
$$y_i = \left(b_i - \sum_{k=1}^{i-1} l_{ik} y_k \right) / l_{ii}. \tag{A.14}$$

(3) Back solution:

$$Ux = y$$

$$y_i = \sum_{k=1}^{i-1} u_{ik} \overset{o}{x_k} + u_{ii} x_i + \sum_{k=i+1}^{n} u_{ik} x_k$$

Thus,
$$x_i = y_i - \sum_{k=i+1}^{n} u_{ik} x_k \quad i = n-1, n-2, \ldots 1. \tag{A.15}$$

A.7 References

1. Whittaker, E.T. and Watson, G.N.: *A Course in Modern Analysis,* Cambridge University Press, Cambridge (1952).
2. Gerschgorin, S.: "Über die Abrenzung der Eigenwerte einer Matrix," *Izv. Akad. Nauk. SSSR* (Ser. Mat. 7, 1931) **16**, 749.
3. Brauer, A.: "Limits for the Characteristic Roots of a Matrix, II," *Duke Math. J.* (1947) **14**, 21.

Index

Accumulation
  matrix, 144
  terms, 176
Adaptive Implicit Method (AIM), 121, 148–149
Alberta, 7, 175
Al-Hussainy and Ramey, 52
Allocation of rates, 154, 156–159
Alternate Diagonal Ordering. *See* Ordering
Alternating Direction Implicit Methods
  Brian's method, 95–96
  comparison to SIP, 112
  Douglas' Crank-Nicolson formulation for 3-D, 96
  Douglas-Rachford for 3-D, 93–95
  failure to coverage, 108
  stability of Peaceman-Rachford ADI, 91–93
  2-D ADI (Peaceman-Rachford), 84–87
  2-D IADI (Peaceman-Rachford), 87–90
Amplification factor, 93, 112
Analytic solution, 5, 66, 115
Anisotropy, 3, 4, 49, 50

Back solution
  in Crout algorithm for band matrices, 100
  in LU decomposition, 26, 199–200
  in SIP algorithm, 108, 111
Back-substitution, 24, 68

Backward difference, 64
Banachiewicz's method, 26
Band
  algorithm, 100–101, 115
  matrices, 98, 133, 147
  width, 99, 101
Basis, 28, 70
Bear, 44
Bernoulli's equation, 45
Berry's near-optimal ordering method, 105–107
Bi-tridiagonal system, 127
Black oil
  flow equations for, 56–59
  reservoir fluid properties, 38, 39
  reservoir simulation, 6, 7
Blair and Weinaug, 140
Block(s)
  active, inactive, 80
  image, 82
  indices, 97
  SOR, 113–115
  tridiagonal, 114
  upstream, downstream, 123, 146
Boundary conditions, 68, 80–82, 83, 115, 190
Brauer, 76, 78, 194–195
Breit and Graue, 170
Brian, 95–96

Bubble-point pressure. *See* Saturation pressure
Buckley-Leverett, 76
Burcik, 40

Camy and Emanuel, 170
Canonical forms, 189–190
Capillary
 hysteresis, 36
 retention, 36
Capillary pressure
 definition, 36
 explicit treatment, 138
 pseudo, 160, 165, 167, 170
 semi-implicit, 140, 141, 173
 zero, 128, 139, 141, 167
Central difference, 64
Centroid of a reference volume, 33
Chappelear and Hirasaki, 172
Characteristic
 equation, 138
 length, 34
 polynomial, 23
 radius, 41
Choleski, 26
Classification
 of partial differential equations, 187
 of reservoir simulators, 6, 7
Closure criteria, 90, 126, 134
Coats, 130, 135, 165, 166, 167, 168, 174, 175
Cofactors, 21
Complex matrix, 195, 196, 198
Compositional
 behavior, 7
 flow equations for, 55–56
 fluid properties, 39–40
 reservoir simulators, 6
Compressibility
 of black-oil fluids, 39
 of rock in terms of porosity, 34
 of rock in terms of water pressure, 130
 zero, 139
Compressible flow
 generalized multiphase flow equation for, 53–57
 single-phase, 51–53
 three-phase flow, 129–135
Computational star, 66, 67, 110
Computer(s)
 cancellation of significant digits, 89
 code, 101
 cost, 148, 159
 high speed digital, 5

Computer(s) *(Cont.)*
 storage, 103–104, 196
 word length in, 74, 75
 work, 103–104
Conservation principle, 2, 3, 5
Constitutive relationship, 34
Constraints on wells, 155, 156–157
Contact angle, 35
Continuity equation
 for multiphase, multicomponent flow, 54
 for single-phase flow, 16, 17, 18, 49
Continuum
 concept of, 33
Convergence
 accelerating, 28, 88, 111, 112–114
 closure criteria for, 90, 126, 134
 definition, 27
 rate, 28, 89, 112–114
Crank-Nicolson method
 applications to single-phase flow in 1-D, 67–69
 applications to single-phase flow in 2-D, 83, 87
 Douglas' application to 3-D ADI for single-phase flow, 96
 noise, 87
 relationship to 2-D Peaceman-Rachford ADI, 92–93
 stability analysis, 73–74, 77
Cross-product, 15
Crout algorithm for band matrices, 100–101
Cycle(s)
 in graph theory, 197, 198
 in iteration parameters, 88, 111, 127

Darcy's law
 for multiphase flow, 54
 for single-phase flow, 45, 46, 47, 49, 58
 Ohm's law analogy, 169–170
Degrees of implicitness, 149, 157
Determinants, 21
Dew point pressure, 40
Diagonal dominance, 100, 117, 198
Diagonal matrices, 19, 114
Difference
 notation, 85
 operator(s), 67, 86
Diffusivity equation, 51
Direct methods, 79, 94, 96, 100
Dirichlet
 boundary conditions, 81–83, 155
 conditions for a Fourier expansion, 72

Discretization
  definition, 63
  error in, 74
  time, 70
Discriminant, 188
Displacement
  in an iteration, 89
  vector, 111, 112
Divergence
  in iterative ADI, 89
  of a vector, 15–16, 29
  theorem, 16
Dot product, 14–15
Douglas, 92, 96. *See also* Douglas and Rachford; Douglas, Peaceman and Rachford
Douglas and Rachford method
  application to three-phase flow, 134–135, 150
  application to two-phase flow, 124–127, 218
  overall equation, 115
  relationship to Peaceman-Rachford ADI, 95
  3-D, single-phase flow, 93–95
Douglas, Peaceman and Rachford, 127
Drainage
  from porous media, 36, 41
  radius, 154
Dufort-Frankel, 76
Dupont, Kendall, Rachford, 111

Edges, in graph theory, 104–107, 196–198
Eigenvalue(s), 23, 30, 71, 73, 138, 191–194
Eigenvector(s), 23, 30, 71, 191–194
Elliptic equations, 80, 96, 108, 189, 190
Equations of state, 47–48
Equipotential surfaces, 13–14
Error
  cancellation, 89, 125
  component, 72, 73, 91
  propagation of, 73
  round-off, 67, 69, 74, 114, 125
  truncation, 61, 63, 67, 74–75, 92–93, 94, 96, 125
  vector, 70
Euclidean space, 13
Euler
  expansions of sine and cosine, 72
  method, 43
Explicit formulation, 65–67
Extended triangle inequality, 193, 194

Factorization
  in LU decomposition, 25, 199
  in the band algorithm, 100
  in the SIP algorithm, 108, 111
Fagin, et al., 172
Fill element, 101, 104–107
First mean value theorem, 164, 186
Flowing bottomhole pressure, 146, 155, 157, 158
Formation volume factors
  definition, 38
  uses in a black-oil simulator, 57
Forward difference, 64
Forward solution
  in LU decomposition, 26, 199–200
  in the band algorithm, 100
  in the SIP algorithm, 108, 111
Fourier('s)
  analysis, 69
  equation, 51, 53
  series, 72
Fractional flow, 76, 139, 158
Fully implicit methods, 121, 141–148, 149
Function(s)
  analytic, regular, 61
  bounded and continuous, 186
  inverse, 188
  nonnegative, 185
  of class C^n, 61–62
  real valued, 185
  vector, 14

Gas
  compressibility, 39
  density, 38
  deviation factor, 38
  equation for a black-oil system, 57, 143
  flow equations, 51–53
  formation volume factor, 38
  ideal, 48, 51–52
  law, 48
  percolation, 153, 173–175
  real or nonideal, 48, 52–53
  reservoir simulators, 6
  solubility, 39
Gauss' theorem, 16
Gaussian elimination, 23–25, 30, 101, 102, 103, 104, 147
  algorithm for tridiagonal matrices, 68–69
Gauss-Jordan reduction, 25
Gauss-Seidel method, 112, 113, 117
Gerschgorin, 76, 78, 192–194

Gradient
  mean fluid, 156
  properties of, 67
  vector, 67
Graph theory, 104–107, 116, 196–198
Gravity force, 162
Grid
  block-centered, 80–82
  lattice-centered, 80–81
  ordering, 97, 101–106
  orientation, 55
  refined, 82, 170

Hadamard, 190
Harmonic analysis, 77. *See also* von Neumann analysis or Fourier analysis
Hearn, 167
Heptadiagonal matrices, 98, 133
Heterogeneous media, 3
History matching, 8–9
Hubbert, 44–45, 54
Hyperbolic equations, 76, 189, 190
Hysteresis effect
  on capillary pressure, 36
  on relative permeability, 37
  on wettability, 35

Identity matrix, 19, 49
Imbibition, 36, 41
IMPES method
  formulation of three-phase problem, 130–135, 150
  formulation of two-phase problem, 128–129
  gas percolation in, 173
  handling of wells in, 157
  stability of, 135–139
  work involved in, 133–134
Incidence matrix, 105
Incompressible flow
  single-phase, 50, 117
  two-phase, 121–129
Inconsistent formulation, 123
Initial conditions, 65, 83, 122, 190
Injection wells, 159
Injectivity index, 146
Inner product, 14–15
Instability, 66–67, 91, 93, 148, 173, 174
Irreducible
  diagonal dominance, 198
  matrices, 117, 196
  saturation, 37, 149
Isotropic media, 49

Iteration matrix, 27
  point Gauss-Seidel, 112
  point Jacobi, 112
  point SOR, 113
Iteration parameters
  for ADI, 88–89
  for SIP, 109, 111–112
  for SOR, 112–113
  optimum in SOR, 113, 117

Jacks, et al., 169, 170
Jacobi, point, 112, 113, 117
Jacobian, 188

Klinkenberg equation, 58
K-value(s), 39–40, 55
Kyte and Berry, 169, 170, 171

Lagrangian method, 43
Laplace's equation, 50
Leap-frog method, 121, 127–128
Leibenzon transformation, 52
Leibniz' rule, 52, 186–187
Leverett J-function, 36
Line successive overrelaxation, 114, 147
Linear algebra, 28
Linear combination, 28, 30, 70
Linearity, 187–188
Linearly dependent vectors, 28, 30
Linearly independent vectors, 28, 30, 70, 191
Loops, 197–198
Lower triangular matrices, 19, 25, 108, 112, 198
LU decomposition, 25–26, 100, 102, 108, 198–200

MacDonald and Coats, 140
Mass
  balance(s), 55
  flux, 16
  fraction(s), 54–55
  unit, 44–45
Material balance equation, 3–4
  volumetric material balance, 56
Mathematical models
  definition of, 2
  material balance equation, 2–4
Matrix
  band, 98, 133, 147
  coefficient, 22, 83, 157
  definition, 18
  eigenvalue problem, 23
  graph of, 104–107, 116, 196–198

Index

Matrix *(Cont.)*
 inverse, 22
 irreducible, 117, 196
 operations, 19–21
 permutation, 195
 polynomial, 192
 rank, 191–192
 reducible, 116–117, 195–196
 spectral radius of, 23
 types, 18–19
Matrix structures, 97, 99–100
Mean fluid gradient, 156
Minors, 21
Mixed boundary conditions, 82
Model(s). *See also* Mathematical models; Physical models
 definition, 2
 immiscible, 7
 miscible, 7
 tank, 3
 thermal, chemical, 7
 zero dimensional, 6
Mole fraction
 overall, 40
 vapor-liquid, 40
Molecular weight, 56
Moles, 56
Momentum equation, 47
Multiphase flow
 Darcy's law for, 54
 generalized equation for, 53–54
 simulators, 121–151
 SIP algorithm for, 112
Multiply connected region, 155

Nabla. *See* Operators
Network analyzer, 4
Neumann boundary conditions, 81–83, 155
Newton's law, 45–46
Nodes, in graph theory, 104–107, 196–198
Nolen and Berry, 140
Nonsymmetric matrices, 29
Normal grid ordering. *See* Ordering
Normalizing factor, 88
Numerical dispersion, 170
Numerical solution, 5

Oil
 compressibility, 39
 density, 38
 equation in a black-oil system, 57, 143
 formation volume factor, 38

Oil *(Cont.)*
 saturated, 39
 undersaturated, 39
Oil-wet, 35
Operator(s)
 del, nabla, 13
 difference, 67, 86
 differential, 13
 for fully implicit formulation, 144–146
 Laplacian, 16
 linear, 187–188
Ordering
 alternate, 116
 alternate diagonal (D-4), 102–104, 116
 diagonal, 116
 near optimal method of Berry, 105–107
 normal, 97, 99
Oscillations
 ADI ringing effect, 87
 from instabilities, 66, 77
Overall equation, 92–93, 94, 96
Overkill, 148, 149

Parabolic equations, 65, 80, 96, 108
Peaceman and Rachford, 83, 87, 91, 93, 94, 95, 115
Pentadiagonal matrix, 83, 98, 114, 133
Permeability
 absolute, 34, 37
 effective, 37
 harmonic average, 124, 170
 principal axes of, 49, 58
 pseudo, 160
 relative, 36–38, 54
 tensor, 47, 49
Permutation matrix, 195
Physical model
 definition of, 2
 network analyzer, 4
 physical reservoir models, 2
 potentiometric model, 4
Point
 Gauss-Seidel, 112
 Jacobi, 112
 SOR, 112–113
Poisson's equation, 50
Polynomial of a matrix, 192
Porosity, 34, 40
Positive definite matrix, 113
Potential
 alternative definition, 45
 gradient of, 13

Potential *(Cont.)*
  Hubbert's, 44–45, 49, 54
  pseudo, 160
Potentiometric model, 4
Price and Coats, 102
Production well, 156–158
Productivity index, 146, 156
Pseudo functions, 153, 160–161
  dynamic, 168–170
  gene-like quality of, 161, 170–171
  VE pseudo functions, 161–168
  well pseudos, 170, 172, 173
PVT data, 8, 38–40

Rank, of a matrix, 190–191
Real gas pseudo pressure function, 52
Recurrence formulas, 69
Reducible matrices, 116–117, 195–196
Reference volume, 33, 41. *See also* Representative elementary volume
Relative permeability, 36–38
  end-point values, 37, 167
  pseudo, 165, 167, 170
  three-phase, 38
Representative elementary volume (rev), 40
Residual(s), 90, 117, 142
Resistance-capacitance networks, 4
Right-hand side expansions, 130–131, 142–144, 176
Ringing effect, 87
Round-off error, 67, 69, 74, 114, 125

Saturation
  definition, 34
  pseudo, 160, 165
Saturation pressure, 39, 175, 176, 177
Scalar product, 14–15
Schilthuis, 3, 5
Second order correct approximations, 64, 85, 145
Sequential method, 141
Settari and Aziz, 82
Sheldon, et al., 128
Simultaneous
  algebraic equations, 23–28
  solution technique for three-phase flow, 141
  solution technique for two-phase flow, 121–127
Single-phase flow
  Darcy's law for, 47, 49
  fluid flow equations, 50–53
Sink, 17, 84, 122, 153, 154, 174
Skew-symmetric matrices, 19, 29

Slightly compressible fluids
  equation of state for, 48
  single-phase flow, 51, 65, 83
Source, 17, 84, 122, 153, 154, 174
Sparse matrices, 98, 101–107, 198
Spectral radius, 23, 27, 79, 113, 193
Stability
  analysis of Crank-Nicolson method, 73–74
  condition for gas percolation, 174
  conditional of IMPES, 138, 139
  of 3-D Douglas-Rachford ADI, 93–94
  of 3-D Peaceman-Rachford ADI, 93
  of 2-D Peaceman-Rachford ADI, 91–92
  matrix method of analysis, 69–71
  nonlinear analysis, 135
  unconditional, 141
  von Neumann analysis of, 71–73, 91–92, 93, 115
Steady-state
  conditions in a reservoir, 96
  flow in wells, 154
Stone('s)
  method for relative permeability, 38
  strongly implicit procedure, 106–112
Stone and Garder, 128
Stright, 169, 176
Strongly coupled, 149
Strongly implicit procedure
  Dupont-Kendall-Rachford method, 111
  for 3-D and multiphase flow, 112, 147
  for 2-D single-phase flow, 107–112
Successive overrelaxation, 112–115, 117, 147
  LSOR, 114
  2LSOR, 114
Surface energy, 35
Symmetric matrices, 19, 26, 29–30, 49

Taylor polynomials, 61, 62, 75, 77
Taylor series, 34, 48, 140
Telegraph equation, 76
Thomas' algorithm, 68–69, 77, 101, 114, 115
Thurnau and Thomas, 176
Transmissibilities
  constant and time-dependent parts, 145
  definitions, 86
  explicit, 123, 128, 138, 139, 173
  implicit, 140
  matrix of, 146
  semi-implicit, 140, 141, 173
  strongly contrasting, 87
  updating of, 140
Trees, 104, 196
Triangle inequality, 193

Tridiagonal
  algorithm, 68–69, 77, 80, 101, 114, 115
  matrix, 68, 83, 84, 88, 99, 133
Truncation error
  analysis of, 74, 75
  in Brian's 3-D ADI, 96
  in 2-D and 3-D Douglas-Rachford ADI, 94
  in 2-D Peaceman-Rachford ADI, 92–93
  local, 61, 63, 67, 125

Underkill, 148, 149
Undersaturated, 39, 175, 176
Unsteady-state flow in wells, 154
Upper triangular matrices, 19, 25, 68, 101, 108, 112, 198
Upstream weighting, 123–124, 146

van Poollen, Breitenbach and Thurnau, 166
Vandermonde matrix, 191
Vectors(s)
  algebra, 14–15
  components of, 12
  definition of, 11, 13
  elementary analysis, 11–18
  error, 70
  gradient, 13
  modulus or norm, 11
  position, 58
  product, 15
  solution, 13
  space, 14, 28, 70
  spanning set, 28
  unit, 12

Vertical equilibrium (VE)
  basic concept, 161–162
  conditions for, 162
  gravity-capillary VE, 162–166
  gravity-segregated VE, 166–167, 176–177
  mathematical conditions, 162
  validation of, 167–168
Viscosity, 46
  pseudo, 160
Viscous force(s), 46, 162
von Neumann
  criterion, 72, 73
  stability analysis, 69, 71, 91–92, 93, 115

Water
  component, 53
  compressibility, 39
  density, 38
  equation, 56, 57, 143
  formation volume factor, 38
Water-wet, 35
Well matrix, 147
Well treatment, 153–158
Wellbore coning, 7
Wellbore index, 146, 154
Well-determined system, 55, 61
Well-posed problems, 65, 83, 190
Wettability, 35
Woods and Khurana, 172–173

Young-Dupre equation, 35

*Better is the end of a thing than
the beginning thereof.*

                              Ecclesiastes